ARCTIC REGION AND ANTARCTICA ISSUES AND RESEARCH

THE UNITED STATES IN THE ARCTIC

KEY AREAS OF STUDY AND ENVIRONMENTAL CHANGE CONSIDERATIONS

ARCTIC REGION AND ANTARCTICA ISSUES AND RESEARCH

Additional books in this series can be found on Nova's website under the Series tab.

Additional E-books in this series can be found on Nova's website under the E-book tab.

ARCTIC REGION AND ANTARCTICA ISSUES AND RESEARCH

THE UNITED STATES IN THE ARCTIC

KEY AREAS OF STUDY AND ENVIRONMENTAL CHANGE CONSIDERATIONS

JOYCE B. STEWART
EDITOR

nova publishers
New York

Library of Congress Cataloging-in-Publication Data

ISBN: 978-1-62618-546-3

Published by Nova Science Publishers, Inc. † New York

CONTENTS

Preface		**vii**
Chapter 1	Arctic Research Plan: FY2013–2017 *Interagency Arctic Research Policy Committee (IARPC)*	**1**
Chapter 2	Changes in the Arctic: Background and Issues for Congress *Ronald O'Rourke*	**89**
Index		**173**

PREFACE

The Arctic region faces growing economic, environmental, and cultural challenges, heightening the need for coordinated research across institutional and national borders. The Arctic environment is undergoing rapid transition as sea and land ice diminish, with tremendous implications for natural environments, human well-being, national security, transportation, and economic development. Notably, the Arctic's central role in environmental change make this region a critical target for research as scientists work to understand the feedback mechanisms influencing global climate, and as policymakers and businesses consider opportunities while minimizing human and environmental costs. This book addresses key areas of study and environmental change considerations facing the Arctic region, with a focus on the 2013-2017 Arctic Research Plan. This plan provides a blueprint for effective Federal coordination of Arctic research for the next half-decade and positions the United States to remain a global leader in Arctic stewardship for many years to come.

Chapter 1 – There is broad scientific consensus that rapid changes in global climate are altering ice and snow cover and affecting Arctic ecosystems, indigenous societies, and natural resources. Research is needed to increase fundamental understanding of these challenges and inform development of sound, science-based solutions. The Interagency Arctic Research Policy Committee (IARPC) is charged with developing five-year plans for Federally sponsored research in the Arctic region. For 2013 to 2017, the IARPC, which consists of representatives from 14 Federal agencies, departments, and offices, has identified seven research areas that will inform national policy and benefit significantly from close interagency coordination. They are:

1. Sea ice and marine ecosystems;
2. Terrestrial ice and ecosystems;
3. Atmospheric studies of surface heat, energy, and mass balances;
4. Observing systems;
5. Regional climate models;
6. Adaptation tools for sustaining communities; and
7. Human health.

These research areas do not encompass all Federal Arctic research activities that will occur over the next five years. Many important investigations outside the scope of this plan will continue to be conducted within individual agencies or through other interagency collaborations.

Chapter 2 – The diminishment of Arctic sea ice has led to increased human activities in the Arctic, and has heightened interest in, and concerns about, the region's future. The United States, by virtue of Alaska, is an Arctic country and has substantial interests in the region. On January 12, 2009, the George W. Bush Administration released a presidential directive, called National Security Presidential Directive 66/Homeland Security Presidential Directive 25 (NSPD 66/HSPD 25), establishing a new U.S. policy for the Arctic region.

Record low extents of Arctic sea ice over the past decade have focused scientific and policy attention on links to global climate change and projected ice-free seasons in the Arctic within decades. These changes have potential consequences for weather in the United States, access to mineral and biological resources in the Arctic, the economies and cultures of peoples in the region, and national security.

The five Arctic coastal states—the United States, Canada, Russia, Norway, and Denmark (of which Greenland is a territory)—are in the process of preparing Arctic territorial claims for submission to the Commission on the Limits of the Continental Shelf. The Russian claim to the enormous underwater Lomonosov Ridge, if accepted, would reportedly grant Russia nearly one-half of the Arctic area. There are also four other unresolved Arctic territorial disputes. The diminishment of Arctic ice could lead in coming years to increased commercial shipping on two trans-Arctic sea routes—the Northern Sea Route and the Northwest Passage. Current international guidelines for ships operating in Arctic waters are being updated. Changes to the Arctic brought about by warming temperatures will likely allow more exploration for oil, gas, and minerals. Warming that causes permafrost to melt could pose challenges to onshore exploration activities. Increased oil and gas exploration and tourism (cruise ships) in the Arctic increase the risk of pollution in the region. Cleaning up oil spills in ice-covered waters will be more difficult than in other areas, primarily because effective strategies have yet to be developed. Large commercial fisheries exist in the Arctic. The United States is currently meeting with other countries regarding the management of Arctic fish stocks. Changes in the Arctic could affect threatened and endangered species. Under the Endangered Species Act, the polar bear was listed as threatened on May 15, 2008. Arctic climate change is also expected to affect the economies, health, and cultures of Arctic indigenous peoples. Two of the Coast Guard's three polar icebreakers—*Polar Star* and *Polar Sea*—have exceeded their intended 30-year service lives, and *Polar Sea* is not operational. The possibility of increased sea traffic through Arctic waters raises an issue concerning Arctic search and rescue capabilities. On May 12, 2011, representatives from the member states of the Arctic Council signed an agreement on cooperation on aeronautical and maritime search and rescue in the Arctic.

The Arctic has increasingly become a subject of discussion among political leaders of the nations in the region. Although there is significant international cooperation on Arctic issues, the Arctic is also increasingly being viewed by some observers as a potential emerging security issue. In varying degrees, the Arctic coastal states have indicated a willingness to establish and maintain a military presence in the high north. U.S. military forces, particularly the Navy and Coast Guard, have begun to pay more attention to the region.

In: The United States in the Arctic
Editor: Joyce B. Stewart

ISBN: 978-1-62618-546-3

Chapter 1

ARCTIC RESEARCH PLAN: FY2013–2017[*]

Interagency Arctic Research Policy Committee (IARPC)

EXECUTIVE OFFICE OF THE PRESIDENT
NATIONAL SCIENCE AND TECHNOLOGY COUNCIL
WASHINGTON, D.C. 20502
February 2013

MEMBERS OF CONGRESS:

I am pleased to forward the five-year Arctic research plan produced by the Interagency Arctic Research Policy Committee (IARPC). Creation of this plan is one of IARPC's responsibilities described in the Arctic Research Policy Act of 1984 (15 U.S.C. § 4108). Fourteen Federal agencies, departments, and offices collaborated to develop this plan, which calls for strong interagency communication, coordination, and collaboration within the framework of the National Science and Technology Council. The IARPC staff also consulted with collaborators in the State of Alaska, local communities, indigenous organizations, non-governmental organizations, and the academic community to ensure that the interests and needs of all stakeholders are addressed appropriately in this research plan.

The Arctic environment is undergoing rapid transition as sea and land ice diminish, with tremendous implications for natural environments, human well-being, national security, transportation, and economic development. The United States and the other Arctic nations require strong, coordinated research efforts to understand and forecast changes in the Arctic. Toward that end, and in furtherance of goals developed by the Arctic Research Commission, this plan focuses on those research activities that would be substantially enhanced by multi-agency collaboration. Many important investigations outside the scope of this plan will continue to be conducted within individual agencies or through other interagency collaborations. I appreciate your support as this Administration works to ensure that the Nation's research efforts in the Arctic are broadly coordinated across the full spectrum of Federal agencies and interests.

---

[*] This report was released by the National Science and Technology Council, February 2013.

Sincerely,
John P. Holdren
Assistant to the President for Science and Technology Director,
Office of Science and Technology Policy

NATIONAL SCIENCE FOUNDATION
4201 WILSON BOULEVARD
ARLINGTON, VIRGINIA 22230
OFFICE OF THE DIRECTOR
February 2013

MEMBERS OF CONGRESS:

The Arctic region faces growing economic, environmental, and cultural challenges, heightening the need for coordinated research across institutional and national borders. Notably, the Arctic's central role in environmental change makes this region a critical target for research as scientists work to understand the feedback mechanisms influencing global climate—and as policymakers and businesses consider opportunities while minimizing human and environmental costs.

To address these challenges and opportunities, and as required by 15 U.S.C. § 4108, the U.S. Interagency Arctic Research Policy Committee (IARPC)—representing 13 Federal agencies, departments, and offices—has crafted this FY 2013-2017 Arctic Research Plan. The plan provides a blueprint for effective Federal coordination of Arctic research for the next half-decade and positions the United States to remain a global leader in Arctic stewardship for many years to come.

This national policy for Arctic research consists of seven overlapping research areas:

1. Sea ice and marine ecosystems
2. Terrestrial ice and ecosystems
3. Atmospheric studies of surface heat, energy, and mass balances
4. Observing systems
5. Regional climate models
6. Adaptation tools for sustaining communities
7. Human health

Each of these research areas poses unique challenges to scientists, but together they promise profound insights into the physical, biological, and human dimensions of the region-insights that will inform decision-makers and guide the Nation as we work to mitigate and adapt to rapidly changing conditions in Arctic communities and around the world.

Importantly, this plan was developed through a coordinated interagency effort that included inputs from Alaskan and other Arctic residents and indigenous leaders. Its implementation will similarly involve a full array of governmental and non-governmental contributors, with the shared goal of developing accurate and understandable science-based information of value to a wide range of users. The plan calls for collaboration across institutional boundaries. This collaboration will be enhanced by leveraging resources, such as the recently launched state-of-the-art research vessel *SIKULIAQ,* which will support and inform the Arctic research community for the next 30 years.

As chair of the IARPC and Director of the National Science Foundation, which has major responsibilities for funding Arctic research, I am committed to working in partnership with our sister agencies and others who are collectively pursuing this important Arctic agenda as a means of achieving a sustainable and secure future for all people around the globe. I thank all of those individuals and groups who have contributed to the completion of this plan and encourage its rapid adoption and implementation.

Sincerely,
Subra Suresh, Director

1. Executive Summary

There is broad scientific consensus that rapid changes in global climate are altering ice and snow cover and affecting Arctic ecosystems, indigenous societies, and natural resources. Research is needed to increase fundamental understanding of these challenges and inform development of sound, science-based solutions. The Interagency Arctic Research Policy Committee (IARPC) is charged with developing five-year plans for Federally sponsored research in the Arctic region. For 2013 to 2017, the IARPC, which consists of representatives from 14 Federal agencies, departments, and offices, has identified seven research areas that will inform national policy and benefit significantly from close interagency coordination. They are:

8. Sea ice and marine ecosystems;
9. Terrestrial ice and ecosystems;
10. Atmospheric studies of surface heat, energy, and mass balances;
11. Observing systems;
12. Regional climate models;
13. Adaptation tools for sustaining communities; and
14. Human health.

These research areas do not encompass all Federal Arctic research activities that will occur over the next five years. Many important investigations outside the scope of this plan will continue to be conducted within individual agencies or through other interagency collaborations.

Sea ice and Marine Ecosystems

Arctic marine ecosystems—driven largely by large-scale changes in sea ice—are moving to new states with the potential for short-term surprises. Over the next five years, the IARPC agencies—including the Department of Defense (DOD), Department of Energy (DOE), Department of the Interior (DOI), National Aeronautics and Space Administration (NASA), National Oceanic and Atmospheric Administration (NOAA), and National Science

Foundation (NSF)—will increase capacity to study such rapid changes through the following activities:

1. Develop a framework of observations and modeling to support forecasting of sea-ice extent on seasonal to annual scales for operational and research needs;
2. Identify and study sites in the Beaufort and Chukchi Seas and the contiguous Arctic Ocean where climate feedbacks are active;
3. Complete deployment of a Distributed Biological Observatory (DBO) in the Arctic Ocean to create long-term data sets on biological, physical, and chemical variability and ecosystem response; and
4. Develop integrated ecosystem-processes research in the Beaufort and Chukchi Seas region.

Terrestrial Ice and Ecosystems

Ongoing changes in the terrestrial Arctic environment that result from climate change are expected to lead to further changes in global climate, or climate "feedbacks," and affect the ability of local communities to adapt. The IARPC has identified the following five priority activities to understand such climate feedbacks and terrestrial ecosystem processeses to be coordinated collaboratively by the DOE, DOI, NASA, NSF, and Smithsonian Institution (SI):

1. Perform glacial-process studies targeting specific dynamic ranges;
2. Coordinate and integrate efforts, including information delivery, that contribute to terrestrial ecosystem research;
3. Identify and study key sites where climate feedbacks are active, including permafrost, snow, hydrates, glaciers, and ice;
4. Investigate the frequency and severity of wildland fires in the Arctic and understand their impacts on vegetation and wildlife; and
5. Conduct socio-economic research to understand ecosystem services as the Arctic tundra changes with increased warming to inform plans for protecting, managing, and adapting to a fragile and changing Arctic environment.

Atmospheric Studies of Surface Heat, Energy, and Mass Balances

Variability in surface-air temperatures—from year-to-year or longer—tends to be larger in the Arctic than in other parts of the globe. Compared with those at low latitudes, atmospheric processes in the Arctic are influenced by unique features—such as polar night, high reflectivity of snow and ice cover, and atmospheric stability—that influence the degree to which aerosols and clouds warm or cool the region. Scientific uncertainties about these unique features must be clarified in order to more fully understand the Arctic atmosphere and its processes. Coordinated remote-sensing and *in-situ* observations, improved representation of atmospheric processes in models, quantification of uncertainty in model outputs, and long-term observational data sets will be critical to addressing scientific uncertainties. DOE, NASA, NOAA, and NSF will collaborate on three activities to support this research area:

1. Improve understanding of short-lived climate forcers (SLCFs) and their role in Arctic amplification through satellite observations, long-term *in-situ* observations, and improved modeling;
2. Improve understanding of processes controlling formation, longevity, and physical properties of Arctic clouds, including the effects of—and sensitivities to—aerosols; and
3. Develop an integrated understanding of Arctic atmospheric processes, their impact on the surface-energy budget, and their linkages with oceanic, terrestrial, and cryospheric systems through improved satellite capabilities, ground-based observations, and representations of Arctic systems in climate and weather-prediction models.

Observing Systems

Arctic change is occurring on multiple spatial and temporal scales. Over the next five years, the DOE, DOI, Environmental Protection Agency (EPA), NASA, NOAA, NSF, Office of Naval Research (DOD-ONR), and United States Coast Guard (DHS-USGC) will focus on nine activities to maintain and strengthen an integrated national and international Arctic observing system to obtain data and information from multiple scales:

1. Facilitate observing-system design for the Arctic;
2. Assess local-resident priorities with respect to climate;
3. Combine *in-situ* and remotely sensed observation of sea ice with local community and traditional knowledge;
4. Conduct long-term monitoring of key outlet glaciers and tidewater glaciers;
5. Monitor the biological and physical state of the Arctic marine environment;
6. Assess the effects of clouds and atmospheric constituents on surface radiation balance;
7. Assess the impact of terrestrial warming and permafrost thawing on the carbon cycle;
8. Improve data access; and
9. Engage indigenous observers and communities in monitoring environmental parameters.

Regional Climate Models

Models of Earth's climate are mathematical tools for understanding current climate processes and projecting future climate variability and change. Improved models of the Arctic region will enhance scientific understanding of processes occurring today, improve the accuracy of projections of future change, guide the design of more effective Arctic field-research campaigns, and support better-informed decision making. The DOE, NOAA, DOI, and NSF will integrate modeling and process-science research through seven activities to improve modeling of Arctic systems and processes:

1. Inventory existing Federal Arctic modeling activities;
2. Encourage coordination to better represent Arctic processes in Earth-system models;
3. Build Arctic-region models to couple with regional and global approaches;
4. Develop models of Arctic land ice mass loss, connections to ocean and atmospheric variability, and implications for sea level;
5. Increase Arctic-model resolution to improve prediction and inform future research and observations;
6. Use model-derived insights to inform process research and vice versa; and
7. Improve understanding of the principle drivers and uncertainties of Arctic climate changes through model validation and verification.

Adaptation Tools for Sustaining Communities

Arctic residents are adapting to new conditions created by rapid environmental change and diverse socio-economic stressors. Over the next five years, the Department of Agriculture (USDA), DOI, Department of State (DOS), EPA, NOAA, NSF, and the SI will assess the resilience and vulnerabilities of Arctic communities to the impacts of climate change. That assessment will help provide residents, community leaders, and policy makers with the knowledge needed to develop sound strategies for successful adaptation. The effort will focus on four activities:

1. In collaboration with local communities, develop methods for assessing community sustain-ability and resilience and determine the efficacy of current adaptation strategies and means for identifying unintended positive and negative outcomes;
2. Identify current vulnerabilities of Arctic communities and ecosystems to climate change and explore the interaction of climate vulnerabilities with socio-economic and other stressors;
3. Develop projections of future climate scenarios and demographic conditions to forecast potential strengths and weaknesses of Arctic human and ecological systems; and
4. Design new research, education, and outreach tools and processes to assist Arctic communities in language and heritage preservation and in cultural revitalization efforts.

Human Health

Arctic indigenous peoples have shorter life expectancies and greater infant mortality rates than their respective national populations. In addition to higher death rates for unintentional injury and suicide, native peoples experience a high prevalence of both infectious diseases and health impacts associated with exposures to environmental pollutants, rapid economic change, and climate change.

IARPC human-health research activities planned for the next five years reflect the priorities of both the Arctic Research Commission *Goals and Objectives for Arctic Research 2009-2010* and the 2011 Arctic Health Ministers meeting held in Nuuk, Greenland. The Department of Health and Human Services, Centers for Disease Control and Prevention (HHS-CDC), EPA, Indian Health Services (HHS-IHS), National Institutes of Health (HHS-NIH), and the U.S. Arctic Research Commission will focus on four activity areas:

1. Continue to expand circumpolar surveillance and research for infectious, non-communicable diseases, trauma, injury, sanitation services, and indoor air quality to help prevent morbidity and mortality;
2. Continue interagency collaboration to monitor the impacts of climate change and environ- mental contaminants on human health and wildlife;
3. Continue to support investigator-initiated research in major health priority areas such as mental health including substance abuse and suicide, obesity, diabetes, and cancer; and
4. Continue to engage indigenous communities and tribal groups in research activities and projects in the Arctic.

2. Introduction and Background

Lead authors:
Brendan P. Kelly, Office of Science and Technology Policy
C. Nikoosh Carlo, National Science Foundation

Introduction

Meeting the Nation's economic, scientific, and environmental needs in the Arctic requires research across diverse disciplines and the involvement of multiple Federal agencies. The Interagency Arctic Research Policy Committee (IARPC), which consists of principals from 14 agencies, departments, and offices across the Federal government, is charged with developing five-year plans for federally sponsored research in (and about) the Arctic region (Arctic Research Policy Act of 1984; Title I of P.L. 98-373 of July 31, 1984). Federal agencies that participate in the IARPC have diverse roles in carrying out the National Arctic Research Policy, which is articulated in National Security Presidential Directive 66/Homeland Security Presidential Directive 25. It mandates that—with respect to research—the IARPC will:

- Meet national security and homeland security needs relevant to the Arctic region;
- Protect the Arctic environment and conserve its biological resources;
- Ensure environmentally sustainable natural resource management and economic development in the region;

- Strengthen institutional cooperation among the eight Arctic nations (the United States, Canada, Denmark, Finland, Iceland, Norway, the Russian Federation, and Sweden);
- Involve the Arctic's indigenous communities in decisions that affect them; and
- Enhance scientific monitoring of and research on local, regional, and global environmental issues.

To ensure that these policy objectives are informed by the best possible science, this plan sets seven priority research areas for the next five years that stand to benefit significantly from close interagency collaboration. This plan does not include all Arctic research activities occurring across the Federal government—many important investigations will continue to be conducted within individual agencies or through other interagency collaborations. Individual agencies have described their own Arctic research priorities in a series of recent reports including:

The U.S. Arctic Research Commission's *Report on Goals & Objectives for Arctic Research 2009 – 2010* (http://www.arctic.gov/publications/2009-10 usarc goals.html);

The U. S. Navy's *Arctic Road Map* (http://www.dtic.mil/cgi-bin/GetTRDoc?Location=U2&doc=GetTRDoc. pdf&AD=ADA516591);

The National Oceanic and Atmospheric Administration's *Arctic Vision and Strategy* (http://www.arctic. noaa.gov/docs/NOAAArctic_V_S_2011.pdf);

The U.S. Geological Survey's (Department of the Interior) An Evaluation of the Science Needs to Inform Decisions on Outer Continental Shelf Energy Development in the Chukchi and Beaufort Seas, Alaska (http://pubs.usgs.gov/circ/1370/pdf/circ1370.pdf); and

The Changing Conditions in the Arctic section of the National Ocean Policy Implementation Plan (http://www.whitehouse.gov/administration/eop /oceans/objectives).

This five-year plan also does not describe the considerable important and complementary Arctic research being conducted outside of the Federal government—by academic, state, tribal, and non-governmental researchers.

Under the auspices of the Arctic Council', Federal agencies also pursue Arctic research through work-ing groups and *ad hoc* task forces. This collaborative work focuses on a wide range of issues, including climate change, indigenous land use, and communications technology.

Successful implementation of this five-year research plan will require close coordination with all of the above listed efforts, the State of Alaska, indigenous organizations, academic institutions, non-governmental organizations, the Arctic Council, and international partners.

Scientists widely agree that the societal and environmental consequences of rapid environmental change are the most pressing scientific concerns in the Arctic region today. Diminishing sea-ice cover is expected to impact the global climate; diminishing ice sheets and glaciers are resulting in sea-level rise; and thawing permafrost is having impacts on both local infrastructure and the concentration of greenhouse gases in the atmosphere. Addressing these challenges will require a sharp research focus on the changing cryosphere—the world's solid-state water, including sea ice, glaciers, snow cover, and permafrost—and its effects on the physical environment, ecosystems, and communities in the Arctic and elsewhere. Key research questions include:

Sea Ice and Arctic Ocean Ecosystems

- At what rates will Arctic sea ice diminish over the next 100 years?
- What will be the consequences of diminishing sea ice for Arctic ecosystems and their inhabitants?
- What will be the consequences of diminishing sea ice for global climate and environments?
- How will Arctic Ocean acidity change in coming decades?
- What will be the consequences of acidification for Arctic ecosystems and their inhabitants?

Ice Sheets and Glaciers

- At what rates will Arctic glaciers and ice sheets diminish over the next 100 years and what processes and forcings are driving the loss?
- What will be the consequences of diminishing glaciers and ice sheets for Arctic ecosystems and their inhabitants?
- How will diminishing glaciers and ice sheets impact global climate and sea level?

Permafrost

- At what rates will Arctic permafrost diminish over the next 100 years?
- What will be the consequences of diminishing permafrost for Arctic ecosystems and their inhabitants?
- How will changes in permafrost impact the global climate system?

These key questions are being addressed by several Federal and non-Federal research efforts including.

The U.S. Global Change Research Program (http://www.globalchange.gov);
The National Ocean Policy Implementation Plan, *Changing Conditions in the Arctic* (http://www.white-house.gov/administration/eop/oceans /objectives); and
The interagency Study of Environmental Arctic Change (SEARCH) (http://www.arcus.org /search/searchscience).

This five-year plan identifies seven priority research areas where interagency cooperation will strengthen and enhance this ongoing work. They are:

- Sea ice and marine ecosystems;
- Terrestrial ice and ecosystems;
- Atmospheric studies of surface heat, energy, and mass balances;
- Observing systems;
- Regional climate models;
- Adaptation tools for sustaining communities; and
- Human health.

Background

The Arctic Research and Policy Act of 1984 (ARPA) established the United States Arctic Research Commission (USARC) to promote Arctic research and recommend a research policy for the region. It also established the IARPC to develop national Arctic research policy and a five-year implementation plan. The IARPC is chaired by the director of the (NSF) and consists of principal representatives from 14 Federal agencies, departments, and offices. The IARPC staff meet monthly and the Principals meet twice per year.

In May 2010, recognizing the increasing participation of multiple agencies in Arctic research, President Obama directed the IARPC to be chartered as a subcommittee of the National Science and Technology Committee. This report constitutes the committee's first deliverable since that change and subsequent revitalization under the leadership of both the Office of Science and Technology Policy and NSF.

The IARPC planning efforts adhere to Section 112 of the ARPA, which defines the Arctic as "all United States and foreign territory north of the Arctic Circle and all United States territory north and west of the boundary formed by the Porcupine, Yukon, and Kuskokwim Rivers (in Alaska); all contiguous seas, including the Arctic Ocean and the Beaufort, Bering, and Chukchi Seas; and the Aleutian chain." While fully accepting this definition, the Committee also emphasizes that the Arctic is part of a larger, changing global system—the boundaries of which must be considered flexible in order to properly study the Arctic's role in important global processes.

Previous Work of the IARPC

Between 1987 and 2007, the IARPC produced 21 volumes of the biannual journal, *Arctic Research of the United States*, which reported Arctic research and results emerging from agencies and partners. The journal's broad vision was to support scientific and engineering research that implements national policy objectives, including:

- Protecting the Arctic environment and conserving its living resources;
- Promoting environmentally sustainable natural resource management and economic development in the region;
- Strengthening institutions for cooperation among the eight Arctic nations;
- Involving indigenous Arctic peoples in decisions that affect them;
- Enhancing scientific monitoring and research on local, regional, and environmental issues (including their assessment); and
- Meeting post-Cold-War national security and defense needs.

Development of the Five-Year Arctic Research Plan

This plan was developed by the IARPC staff: Jonathan Berkson, USGC; Shella Biallas, DOI; John Calder, NOAA; C. Nikoosh Carlo, NSF; Ashley Chappell, NOAA; Kathy Crane, NOAA; Richard Eckman, NASA; Wanda Ferrell, DOE; William Fitzhugh, SI; Martin O. Jeffries, DOD; Brendan P. Kelly, OSTP; Igor Krupnik, SI; Michael Kuperberg, DOE; Marya Levintova, NIH; Kim Mcgraw, DOI; Adrianna Muir, DOS; Alan Parkinson, CDC; James Partain, NOAA; Robert Sanford, NSF; Sandy Starkweather, NOAA; Simon Stephenson, NSF; Louis Tupas, DOA; Taneil Uttal, NOAA; Thomas Wagner, NASA. The Arctic Research Commission also provided valuable input to this plan.

Evolution of Scientific Studies in the Arctic

Early scientific information about the Arctic was primarily geographic and collected by 18th and 19th century European explorers (Beechey 1831; Hall 1866). For most of the following 100 years, scientific information came mainly from scattered efforts in ethnography and natural history, mostly associated with expeditions by the U.S. Signal Corps and others (Dall 1870; Ray 1885) and with management of fishing and hunting (Allen 1880; Elliot 1898).

The first intensive investigation along the Arctic coast of Alaska took place with the International Polar Year (1881–1884) when the Signal Corps occupied a research station at Barrow, Alaska (Baker 1982). In the 1940s, ONR established the Naval Arctic Research Laboratory at Barrow, and studies of Arctic environments have been carried out there almost continuously under the administration of the Navy, the University of Alaska, and the North Slope Borough.

In the late 1950s, the Federal government considered detonating nuclear devices to create a port along the Chukchi coast of Alaska (AEC 1959). The Atomic Energy Commission contracted an investigation of potential environmental impacts and, thereby, provided baseline information about the Chukchi coast of Alaska and its near-shore waters (Willimousky and Wolfe 1966). In the 1970s and 1980s, an interagency agreement between the DOI's Bureau of Land Management and the Department of Commerce (DOC) NOAA created the Outer Continental Shelf Environmental Assessment Program to study the potential impacts of offshore oil development in sub-Arctic and Arctic Alaskan waters. For a decade or more, hundreds of studies looked at ice movements and deformation, mammals, birds, fish, benthos, plankton, microbiology, chemistry, oceanography, meteorology, and geology. Some of the work was eventually published in peer-reviewed literature, and all of it has been assembled by the Alaska Resources Library and Information Services housed on University of Alaska Anchorage campus (http://www.arlis.org/ resources/ocscap-reports).

Through the 1980s, most Arctic research was conducted in isolated disciplines such as biology, geology, and anthropology. Toward the end of the decade, systems science matured (Ashby 1956; von Bertalanffy 1972; Lawton 2001), and a realization that human activity was driving rapid change in the Arctic began to prompt interdisciplinary work. The importance of interactions among systems—atmosphere, geosphere, hydrosphere, and biosphere—was underscored by the International GeosphereBiosphere Programme and the World Climate Research Programme. The notion also stimulated the formation of NSF's Arctic System Science Program, with a research focus on paleoenvironments and contemporary studies of interactions among ocean, land, and atmosphere. Eventually, human dimensions were included in studies of contemporary and paleoenvironments.

While the IARPC's mission is focused on the Arctic region, many new systems science approaches have been developed to consider high latitude phenomena (especially those associated with ice sheets, sea ice, and atmospheric coupling) in a broader perspective. As a result, there have been many commonalities in the evolution of Arctic and Antarctic science.

In 1997, scientists from 25 institutions called for a coordinated effort to understand rapid environmental change in the Arctic. Their efforts led to the formation of the SEARCH (http://www.arcus.org/search/ index.php). They produced science and research plans based on three main components: observing, understanding, and responding to Arctic change. The Arctic Observing Network (AON; http://www.arcus.org/search/aon)—a component of SEARCH—aims to track and foster understanding of the complex, rapid environmental changes taking place in the Arctic through modeling, reconstructions of paleoen-vironments, and process studies of the environment, socioeconomics, cultures, and human health. Responding to change requires consideration of possible adaptive responses of Arctic communities and possible effects on people living outside the Arctic region. Rapid warming of the Arctic has led to dramatic declines in sea-ice extent and thickness with local and global impacts. Arctic sea ice influences atmospheric circulation patterns and precipitation as far south as the tropics (Budikova 2009). Similarly, warming has led to substantial losses of land ice, primarily from Greenland and Alaska, which are now responsible for a substantial fraction of observed sea level rise.

Summary

Interagency cooperation on Arctic research is more important than ever. Rapid changes are affecting the region's biota and people in many ways, including by increasing access to the region for energy and mineral development, shipping, tourism, and military operations—human activities that may carry both risks and opportunities for the Arctic region. Federal agencies are conducting scientific research to fundamentally understand those changes, risks, and opportunities. Policymakers are increasingly relying on that science to make decisions and form practical responses. This IARPC research plan aims to support those decisions with enhanced interagency cooperation on Arctic research to address the most pressing science needs.

References

Allen, J. S. (1880). History of North American pinnipeds; a monograph of the walruses, sea-lions, sea-bears and seals of North America. Washington: Government Printing Press.

Ashby, W. R. (1956). Introduction to Cybernetics. New York: Wiley.

Atomic Energy Commission (AEC). (1959), Roads and Core Holes: Project Chariot Site Plan. Holmes and Narver, Inc.

Baker, F. W. G. (1982). The First International Polar Year, 1882-1883. Polar Record, 21(132), 275-285.

Beechey, F. W. (1831). Narrative of a voyage to the Pacific and Beering's Strait. London: Henry Colburn and Richard Bently.

Budikova, D. (2009). Role of Arctic sea ice in global atmospheric circulation: A review. Global and Planetary Change 68, 149-163.

Dall, W. H. (1870). Alaska and its resources. Boston. Lee and Shepard.

Elliot, H. W. (1898). Report on the seal islands of Alaska. House Exec. Doc. 92, pt. 3-1.

Hall, C. F. (1866). Arctic researches and life among the Esquimuax: being the narrative of and expedition in search of Sir John Franklin, in the years 1860, 1861, and 1862. New York, NY: Harper and Bros., p. 565

Lawton, J. (2001). Earth System Science. Science 292:1965.

Ray, P. H. (Ed.). (1885). Report of the International Polar Expedition to Point Barrow, Alaska. 48th Congress, 2nd Session, House Exec. Doc. 44.

Steller, G. W. (1988). Journal of a Voyage with Bering 1741-1742 O. W. Frost (Ed.), Stanford University Press.

von Bertalannfy, L. (1972). The history and status of general systems theory. The Academy ofManagement Journal 15, 407-426.

Wilimousky, N. J., & J. N. Wolfe. (1966). Environment of the Cape Thompson Region. Alaska: U.S. Atomic Energy Commission, Washington, D.C.

3. Research Areas

3.1. Understand Sea-Ice Processes, Ecosystem Processes, Ecosystem Services, and Climate Feedbacks in the Beaufort and Chukchi Seas and the Contiguous Arctic Ocean

Lead Author:
John Calder, NOAA
Agency Partners:
DOD, DOI, NASA, NOAA, NSF

The annual Arctic Report Card (http://www.arctic.noaa.gov/reportcard) integrates the latest information on the state of the Arctic based on input from more than 100 authors around the world.

The 2011 Report Card concludes that data collected since 2006 are sufficient to indicate a shift in the Arctic Ocean system that is characterized by the persistent decline in the thickness and summer extent of sea-ice cover and a warmer, fresher, and more acidic upper ocean. There is, moreover, growing evidence that those changes are forcing marine ecosystems in the Beaufort and Chukchi Seas and the contiguous Arctic Ocean toward new and generally unknown states, with the potential for short-term surprises (Krupnik and Bogoslovskaya 1999; Grebmeier 2006; Gradinger, *et al.* 2010).

The IARPC agencies will work together over the next five years to enhance understanding of changing sea-ice-ocean-atmosphere interactions and feedbacks; improve sea-ice forecasting at various spatial and temporal scales; and detect and understand ecosystem change.

Both "understanding" and "fore-casting" involve the application of models, whose outputs will inform decision making and planning for future observations. In conjunction with work described in other sections of this plan (e.g., section 3.2), agency partners will enhance efforts to organize, disseminate, and analyze relevant data and informa-tion.

Applying modern cyberinfrastructure will improve capabilities for integrating data from different sources and blending different types of data to produce new insights and information.

These advances are needed to be as responsive as possible to the needs of local residents, businesses, government decision-makers, and managers of Arctic Ocean resources.

The work focuses on four activities.

3.1.1. Develop a Framework of Observations and Modeling to Support Forecasting of Sea-Ice Extent on Seasonal to Annual Scales for Operational and Research Needs

Why Do This?

The Arctic Ocean system has shifted to a new state. Evidence cited in the Arctic Report Card shows that the minimum extent of summer Arctic sea ice from 2007 to the present has fallen below the previously established trend line for the period 1979 through 2006, and the rate of sea-ice loss exceeds that projected by coupled climate models (Stroeve *et al.* 2007)[2].

Continued loss of sea ice will have important consequences for marine and terrestrial ecosystems, coastal communities, maritime transportation, natural resource development, regional and global weather and climate, and national security. Understanding and predicting the consequences of continuing sea-ice loss on the marine ecosystem will require better understanding of the Arctic Ocean environment and processes to improve sea-ice forecasts and predictions at a variety of spatial and temporal scales.

Timeframe

Mid-term (3-5 years)

Expected Outcomes

- Improved operational and research forecasts/projections to support safe operations and ecosystem stewardship on a seasonal to annual basis.
- Reduce uncertainty in predictions and projections at longer time scales and over the entire Arctic marine area, for better informed policy and decision making at local, state, national, and international levels.
- Significantly improved seasonal weather and sea-ice models and forecasts to fill a critical gap in marine weather and climate services, since at seasonal and shorter time scales, sea-ice and weather forecasts are tightly coupled and must be pursued together. Improved models and forecasts will benefit community and subsistence activities, management of protected marine resources (including ice-dependent species), marine navigation, and industry operations.
- At annual and longer time scales, reduced uncertainty and increased accuracy of sea-ice projections and enhanced understanding of how newly sea-ice-free areas influence weather and climate, not just in the Arctic but in the global system. Enhanced understanding will lead to sustainable infrastructure and community planning and aid in projection of regional and global climate impacts forced by changes in the Arctic.

Milestones

- Convene interagency expert group on sea-ice forecasting to develop multi-year implementation plan, coordinate on-going observation and modeling, and determine needed improvements to reduce uncertainty in forecasts [DOI's Bureau of Ocean Energy Management (BOEM), NASA, NOAA, NSF, ONR; FY2013][3].
- Engage with stakeholders and users to determine needs for sea-ice forecasts and products through venues such as the Alaska Marine Science Symposium and Alaska Forum on the Environment (BOEM, NOAA, NSF, ONR; FY2013).

- Continue the Sea Ice Outlook and Sea Ice for Walrus Outlook to evaluate diverse sea-ice forecasting approaches and fill a valuable user need (NSF, NOAA; FY2013).
- Launch Ice, Cloud, and land Elevation Satellite (ICESat) 2 satellite altimetry mission to continue the record of sea-ice thickness measurements and land ice elevation change (NASA; FY2016).
- Launch Gravity Recovery and Climate Experiment (GRACE) Follow-On satellite mission to continue the record of changes in Arctic Ocean circulation and land ice mass loss (NASA; FY2017).
- Develop algorithms for the Advanced Microwave Scanning Radiometer 2 (AMSR2), recently launched on Japan's GCOM-W, to continue and enhance passive microwave record of sea-ice extent (NASA, NOAA; FY2014).
- Improve knowledge of sea-ice melting through various activities such as ONR's marginal ice zone program and NASA's Operation IceBridge mission (NASA, NOAA, NSF, ONR; FY2017).

Science and Technology Gaps and Needs

Enhanced environmental observations and improved modeling capabilities are needed to meet the requirements for improved sea-ice forecasts on a seasonal to annual basis. Many of the observations and modeling capabilities would also be valuable for sea-ice forecasts at sub-seasonal time scales. Appropriate collaborations are needed to ensure data and model outputs are available to support forecasting at both time scales.

Both *in-situ* and remotely sensed observations will be needed, taking full advantage of international remote sensing assets. Processes requiring assessment include ice growth, export, melt, and albedo change, with *in-situ* measurements coupled to remotely sensed observations to create a pan-Arctic set of sea-ice state variables for data assimilation and model initialization.

Specific sea-ice characteristics that need to be assessed are ice concentration, ice thickness, snow thickness, ice type (first-year vs. perennial), ice motion, leads/polynyas, melt pond fraction, surface albedo, temperature, and bottom and top ablation. Ocean mixed layer temperature, tidal, bathymetric, and circulation data are also needed. Most of these data also have applications to ecosystem and coastal zone studies and would lead to improved understanding of storm surges.

Planning for observations must include input from modeling centers to ensure data collection in a manner appropriate to needs for initialization, validation, and assimilation of various forecast models. Continuous or frequently repeated data collection will be needed, including broad surveys of ice conditions in at least spring and fall, to initialize forecasts of both ice loss and regrowth. To meet the observational requirements in a cost-effective way, it will be necessary to take full advantage of all available observing platforms (e.g., ships, aircraft, fixed offshore platforms, coastal locations) on an opportunistic basis. Partnerships with national and international organizations and with private industry are needed so that platforms can be equipped with instrumentation for many of the needed observations on a mutually beneficial basis. Whenever possible, data should be returned in near real-time to support forecasting at shorter time scales and to verify sensor performance.

International collaboration will be a necessary component of this project, not only because Canada and Russia share the target region with the United States, but also because international collaboration is needed for access to critical remote sensing data. The most important of these are the international satellite radar missions, which are important for constraining sea-ice age, thickness, and motion [e.g., European Space Agency (ESA) CryoSat-2 and Canadian Search And Rescue Satellite Aided Tracking (SARSAT) Missions].

3.1.2. Identify and Study Sites in the Beaufort and Chukchi Seas and the Contiguous Arctic Ocean Where Climate Feedbacks Are Active

Why Do This?

There are several feedback processes active in the Arctic marine environment, including those related to albedo and radiative-balance changes and air-sea fluxes of heat, moisture, and greenhouse gases. The loss of sea ice can affect all of these processes. The extent of summer Arctic sea ice is declining, the ratio of single-year to multi-year ice continues to increase, and the rate of sea-ice loss exceeds that projected by coupled climate models (Stroeve *et al.* 2007).

The sea-ice edge during the Arctic spring through fall seasons is dynamic with rapid and large location changes in response to amount of snow cover; clouds, solar radiation, and albedo; winds and ocean waves; and air and water temperatures. Interactions and feedbacks among those variables, from local to regional scales, can amplify Arctic-wide climate change and sea-ice retreat (Perovich and Richter-Menge 2009). Better knowledge about such feedbacks is vital for understanding ice-air-ocean system processes, improving daily to seasonal weather and sea-ice forecast models, and increasing the accuracy of longer-term sea-ice and climate projections.

Changes in sea ice and other portions of the cryosphere in the Beaufort and Chukchi Seas region may affect fluxes of greenhouse gases. Recent work on the Beaufort Sea shelf found no conclusive evidence of massive methane venting from subsurface systems (Coffin *et al.* 2010), and there is no evidence in high latitude atmospheric greenhouse gas concentration data that natural emissions of methane have increased significantly in the last decade (Dlugokencky and Bruhwiler 2012).

On the other hand, recent evidence (Shakova *et al.* 2010) indicates that methane is being released from or through thawing permafrost under shallow coastal seas north of Eurasia. The amount of methane available for release is potentially very large, but there are no data on the current rate of release throughout the Arctic or how the rate might change. The shallow shelves off northern Siberia and around the McKenzie Delta are important marine areas for methane research, and partnerships with Russia and Canada are essential to this work.

Carbon dioxide fluxes between the atmosphere and ocean may change with a persistent loss of sea-ice cover and warming of the Arctic Ocean. Carbon dioxide may be increasing or decreasing in the ocean at varying rates (Bates *et al.* 2011), and predicting the future influence of these fluxes on atmospheric carbon dioxide levels and on ocean acidification is not possible at present.

Timeframe

Mid-term (3-5 years)

Expected Outcomes

- Improved understanding of the role of feedback processes driving sea-ice variability and change; improved sea-ice forecasts to support safe operations, ecosystem stewardship, and improved predictions and projections of regional and global climate; better informed policy and decision making at local, state, national, and international levels.
- Improved daily, weekly, and seasonal weather and sea-ice models and forecasts will fill a critical gap in marine weather and climate services with benefit for community and subsistence activities, management of protected marine resources (including ice-dependent species), marine navigation, and industry operations.
- Reduced uncertainty and increased accuracy of sea-ice projections and understanding of how newly sea-ice-free areas influence weather and climate, which will lead to sustainable infrastructure and community planning and aid in projecting of regional and global climate impacts forced by changes in the Arctic.
- Improved understanding and ability to forecast (based on both *in-situ* and remotely sensed data) Arctic environmental change.
- Improved estimates of the contribution of Arctic greenhouse gas fluxes to climate warming.

Milestones

- Report of a workshop on sea-ice forecasting (NOAA; FY2012[4]).
- Report of a workshop on the future of Arctic sea-ice research and forecasting, National Academy Polar Research Board (NASA, ONR; FY2013).
- Work through the Oil Spill Recovery Institute (OSRI) in conjunction with State of Alaska and response organizations to apply research results to oil spill response planning (DOI, NOAA, USCG[5]; FY2013).
- Investigate the marginal ice zone:
  (1) Emerging Dynamics of the Marginal Ice Zone (ONR; FY2016).
  – Science and field experiment planning; equipment development and testing (ONR; FY2013).
  – Main field experiment in the Beaufort Sea (ONR; FY2014).
  – Data analysis and synthesis (ONR; FY2016).
  (2) Marginal Ice Zone Observations and Processes Experiment (MIZOPEX): (NASA, NOAA; FY2013).
- Develop and test large-class, heavy payload unpiloted aerial systems (UAS) for sea-ice characterization (NASA, NOAA; FY2013-14).
- Investigate sea-state and boundary layer physics in the emerging Arctic Ocean (ONR; FY2017).
  – Science and field experiment planning; equipment development and testing (ONR; FY2014).
  – Main field experiment in the Beaufort and Chukchi seas (ONR; FY2015).
  – Data analysis and synthesis (ONR; FY2017).
- Initiate interagency activity to improve application of remote sensing and buoy/mooring data to sea-ice forecasting (NOAA; FY2012).
- Investigate characterization of the circulation on the continental shelf areas of the Northeast Chukchi and Western Beaufort Seas (BOEM; FY2012).

- Continue operation IceBridge acquisition of sea-ice surface elevation and supporting data, and expand Arctic sea-ice observations to constrain melting processes (NASA; FY2017).
- Initiate interagency evaluation of trends and significance of methane flux to the atmosphere in Arctic regions (DOE, NOAA, USGS[6]; FY2015).
- Initiate a dialogue with Roshydromet and Russian Academy of Science on a potential investigation of the current rate of methane release from the shallow shelves off northern Siberia (NOAA; FY2013).
- Initiate a dialogue with Canadian agencies on collaborative methane research along the Beaufort Sea coast (DOE; FY2013).
- Identify optimal sites for short-term process studies underpinned by long-term observations (all agencies; FY2014).

Science and Technology Gaps and Needs

New and enhanced communications technology for *in-situ* orbital and surface-based sensors is needed to provide real-time, integrated observations and products derived from atmosphere, ice, and ocean. The sensors will support investigations of the interactions and feedbacks among variables that control sea-ice concentration, thickness and motion, and the location of the ice edge. Deployment of *in-situ* sensors and support for process studies will require coordination with and access to charter and non-charter vessels capable of operating during spring, summer, and fall. Research into interactions and feedbacks among the atmosphere, ice, and ocean requires improved process models. Improving forecasts and predictions will require models that assimilate advanced observing data and derived products; provide time-varying sea-ice concentration, thickness, and ice-edge location at high temporal and spatial resolution; and fully couple ice-ocean-atmosphere processes. Improved sea-ice predictions also are critical to meeting objectives described in the following parts of this section requiring new science integration efforts on an interdisciplinary basis.

3.1.3. Complete Deployment of a Distributed Biological Observatory in the Arctic Ocean to Create Long-Term Data Sets on Biological, Physical, and Chemical Variability, Change, and Ecosystem Response

Why Do This?

Changes in location and timing of the seasonal ice edge can have profound effects on sea floor and water column marine ecology and human activity (Grebmeier *et al.* 2012; Huntington 2009). These changes also affect the ability of ice-dependent marine mammals to reproduce and rear young on ice (Kelly 2001; Kelly *et al.* 2010; Cameron *et al.* 2010; Hezel *et al.* 2012). Changes in zooplankton availability can affect distribution and abundance of baleen whales, which are important to subsistence cultures (Ashjian *et al.* 2010; Moore *et al.* 2010). Likewise, stranding of ice-dependent species on land likely reduces their survival or reproductive rates and will change their availability to subsistence hunters. Relationships among ice-edge retreat, changes in plankton dynamics, loss of summer sea ice, and foraging success of whales and ice-dependent species are poorly understood (Moore and Huntington 2008; Kovacs *et al.* 2011), as are the effects of these changes on Alaska Natives who depend upon such species (Metcalf and Robards 2008). In the initial study region (Fig. 3.1.1) changes in ecosystems will be partly driven by the varying flux of mass, heat, salt, and nutrients through the Bering Strait as well as by sea-ice loss and other changes in the physical state of the Arctic Ocean. Also, the effects on marine ecosystems of ocean acidification, happening now and projected to be greater in Arctic waters than anywhere else, are largely unknown.

The Distributed Biological Observatory (DBO) (http://www.arctic. noaa.gov/dbo) integrates biological and physical sampling from both moorings and ships using a collaborative network of logistical sup-port (Grebmeier *et al.* 2010). Remote sensing and advanced *in-situ* technologies will be critical to the observatory's success. Information from the DBO will provide a better understanding of how climate change affects Arctic biology and what steps will be necessary to improve stewardship of the Arctic marine ecosystem as human use and economic development increase. The DBO is intended to span multiple decades and likely will evolve as new approaches and technologies become available.

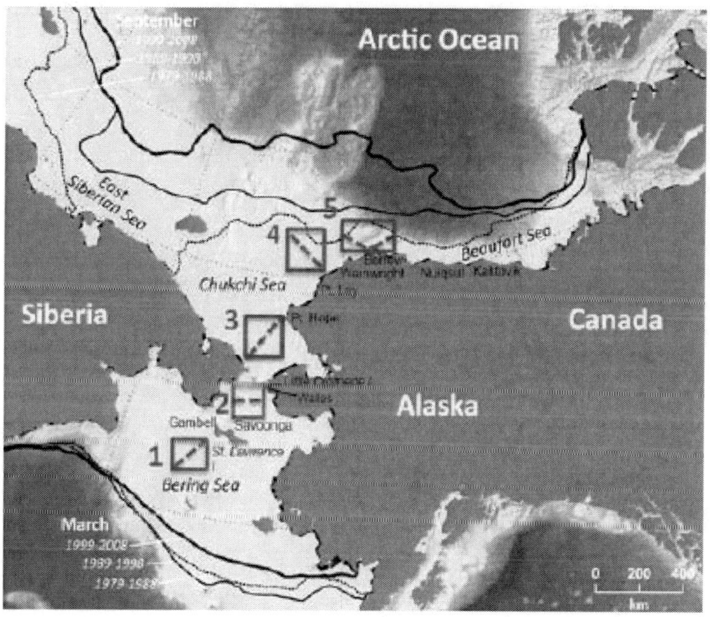

Figure 3.1.1. Five possible regional locations of Distributed Biological Observatory transect lines and stations for standard hydrological and biological measurements in the Pacific Arctic sector. These locations were selected because they are known biological "hot spots" where change might be easier to detect.

Relevant U.S. agencies (including NOAA, NSF, and NASA) coordinated through the IARPC and international partners (including Canada, China, Japan, Korea, and Russia) coordinated through the Pacific Arctic Group (PAG)[7] will take advantage of current research programs to create a marine DBO in the region for consistent, long-term monitoring of biophysical responses in pivotal oceanographic areas along a north-south latitude (Figure 3.1.1). Each area exhibits high biological productivity, biodiversity, gradients in ecosystem properties, and direct linkages to subsistence-based coastal communi-ties. All areas are projected to experience increased commercial use with the loss of sea ice. As sea ice retreats, the DBO will track the rate of ecosystem change and identify impacts. The DBO will provide critical information on the biodiversity of this region and a baseline for assessing how biodiversity will respond to climate change and loss of sea ice. The DBO will also provide baseline information necessary to assess and mitigate potential impacts of offshore resource development on subsistence activities. Observations will be conducted in collaboration with international partners and coordi-nated through the Pacific Arctic Group, which is a forum for discussion of science issues relevant to the Pacific sector of the Arctic. Its participants are drawn primarily from Canada, China, Japan, Korea, Russia, and the United States.

Timeframe

Pilot-project and planning (3-5 years); decadal-term implementation

Expected Outcomes

The DBO will:

- Provide a knowledge-resource base to improve the ability of resource management agencies (e.g., BOEM and NOAA) to determine the effects of their actions on marine resources, resulting in improved conservation, protection, and management of Arctic coastal and ocean resources.
- Improve understanding of the effects of Arctic ecosystem and climate changes on subsistence cultures in the region.
- Provide information on variability and change in the physical environment at DBO sites and the associated response of planktonic, pelagic, and benthic communities to enable improved management and use of marine resources, including subsistence use of marine mammals.
- Develop or improve integrated physical-ecosystem models that relate past variability and change in both physical and biological conditions at DBO sites and inform resource managers of potential future conditions under different scenarios.
- Aid in developing risk-averse strategies to maximize the resilience of marine ecosystems, and develop strategies to mitigate and adapt to adverse impacts.
- Increase understanding of the ecological implications of: – Increasingly early ice-edge retreat and absence of summer sea ice; – Increased severity of storms during the ice-free season;
  - Ice-dependent species forced to spend time on land, including impacts of human bance; and
  - Ocean acidification on Arctic marine ecosystems, especially plankton and calcareous se - floor dwelling organisms important as prey to subsistence species.
- Increase ability to monitor and assess environmental conditions under changing climate narios through new collaborations and partnerships among participants of the Pacific Group and across U.S. agencies.
- Forge connections between a U.S. observing network and similar networks of other countries.

Milestones

- DBO partners conduct pilot research cruises (http://www.arctic.noaa.gov/dbo/) (NOAA, NSF; FY2013).

- Creation of DBO satellite ocean color record (NASA; FY2014).
- The Arctic Observing Network (AON) subcommittee established by the IARPC organizes the DBO interagency working group to develop U.S. plans and priorities (NASA, NOAA, NSF; FY2012).
- Initiate a dialogue with *Roshydromet* and Russian Academy of Science on developing DBO tions in Russian territory as a complement to those in U.S. waters (NOAA; FY2013).
- Pacific Arctic Group meets annually to review results from 2010-2013 pilot activities (NOAA; FY2013).
- Report in 2014 on International DBO activities and results to date (NOAA; FY2014).
- Updated DBO concept and national/international plan for decadal-scale implementation release in 2014 will include identification of satellite resources that will be critical to the DBO. Ocean color, sea surface temperature (SST), sea surface height (SSH), sea surface (SS) salinity, and winds are all key measurements in the cloudy Arctic for ecosystem characterization (NASA, NOAA, NSF; FY2014).
- Starting in 2015, DBO partners execute decadal-scale plans and prepare periodic assessments on physical and ecological state of Pacific Arctic marine environment using not only DBO data, but also data from BOEM, the North Pacific Research Board, and other sources (BOEM, NOAA, NSF; FY2015).
- Report annually on recent developments at the Alaska Marine Science Symposium and seek coordination with Alaska state agencies, oil industry, and other non-Federal organizations. (Selected agencies as appropriate; FY2014).

Science and Technology Gaps and Needs

The DBO will require agencies to pursue new technologies for continuous, year-round, real-time observations of key physical, chemical, and biological variables; improve coordination with and access to charter and non-charter vessels capable of working in Arctic areas during the spring, summer, and fall; enhance application of satellite- and aircraft-based sensors (including use of unmanned aircraft); and improve use of community based observations and instrumented animals.

3.1.4. Develop Integrated Ecosystem Research in the Beaufort and Chukchi Seas

Why Do This?

Research is needed to identify and understand processes that control ecosystem structure and function as well as their sensitivities to changes in physical and chemical environments. Results from such process studies inform models that project future ecosystem status and provide critical information to support adaptation efforts. Coordination of the Bering Sea Ecosystem Study (NSF) and the Bering Sea Integrated Ecosystem Research Program (North Pacific Research Board) stands as a model for how an interagency study might be organized (Bering Sea Interagency Working Group (2006).

There are numerous outstanding research questions that will benefit from an interagency approach, such as:

- How will ecosystems respond to expected continued warming and acidification of the waters in the Chukchi and Beaufort seas and the contiguous Arctic Ocean?

- How might fluxes of mass, heat, salt, and nutrients from the Bering Strait, from the Mackenzie River, from Siberian coastal currents, or from Atlantic water intrusion impact ecosystems in the study region?
- How will ice-edge ecosystems adapt to changes in location and timing of sea-ice retreat and re-growth?
- Will the trend toward increased areas of open water seen in recent summers lead to seasonal deepening of the mixed layer and alter ecosystem processes?
- Will southern species establish themselves in the Arctic Ocean and might the resulting stocks approach commercial size?
- Will eastern and western species and stocks intermix, and will intermixture decrease diversity and/or adaptation?
- How well do ecosystem models describe the current state of ecosystems in the study region, and what improvements are needed to enable skillful projection of future ecosystem states?
- What new observing tools or technologies are needed to improve understanding of ecosystem processes?
- How will degradation of submerged permafrost and natural gas hydrates impact ecosystems in the Chukchi and Beaufort shelf regions?
- How will increased oil and gas drilling in the Chukchi and Beaufort seas impact natural ecosystems?
- How will shipping traffic and other human activities impact subsistence species and the communities that depend upon them?

Timeframe

Mid-term (3-5 years)

Expected Outcomes

This research will create new knowledge about the regional ecosystem, including its key components and their linkages, and insight into how it responds to perturbation. This information will help in the development of hypotheses about responses to long-term trends such as persistent loss of summer sea ice, ocean warming, and ocean acidification. Scenarios for future subsistence and commercial use of living marine resources could be constructed from the research results and serve as guides for longer-term sustained observations. Results from process studies inform models that can be used to project future ecosystem status and allow proactive adaptation to changing conditions.

Milestones

- Conduct interagency and international workshops and consultations during 2013 to identify high priority research themes and objectives and coordinate funding and logistic plans (Selected agencies as appropriate; FY2013).
- Perform synthesis and assessment during 2013-2014 on existing data and information to provide foundation for new research activities (Selected agencies as appropriate; FY2014).
- Initiate 3-5 year research activities starting in 2014 with interagency/international results integration mechanism (Selected agencies as appropriate; FY2014).

- Demonstrate new and updated cyberinfrastructure tools to enhance data integration and application and identify opportunities for sharing of technology and tools among interagency partners (all agencies; FY2013).
- Conduct initial science integration conference in 2016 (Selected agencies as appropriate; FY2016).
- Conduct environmental and integrated risk assessments to evaluate the potential impacts of oil/natural gas production on ecosystems in the Beaufort Sea (DOE; FY2014).
- Evaluate ecosystem impacts of oil and gas development and potential oil spills (including effects of oil dispersants and a mixture of oil and dispersants) on the early life-history stages of Arctic cod (NOAA; FY2013).

Science and Technology Gaps and Needs

Improved partnerships between scientists and local-knowledge holders are needed as are the means for reviewing and funding such partnerships. A number of successful projects have been supported that encourage or require such partnerships, but there is recognition that more needs to be done to capture local knowledge in the Arctic. Agencies have differing mechanisms for supporting these types of partnerships but are open to good ideas generated at the working level (local residents, academic experts, and agency scientists) and proposed for consideration. The planned Arctic Observing Summit in 2013 is one opportunity for such ideas to be presented. Improved biophysical models will be needed, along with a parallel effort to test and validate the model outputs. Full-year investigations will create new requirements for observing platforms, data storage, and transmission.

References

Ashjian, C. J., Braund S. R., *et al.* (2010). Climate variability, oceanography, bowhead whale distribution and Inupiat subsistence whaling near Barrow, Alaska. *Arctic* 63(2), 179-194.

Bates, N. R., Cai W-J., & Mathis, J. T. (2011). The ocean carbon cycle in the western Arctic Ocean: Distribution and air sea fluxes of carbon dioxide. *Oceanography* 24(3), 186-201.

Bering Sea Interagency Working Group. (2006). Climate change and the Bering Sea ecosystem: An integrated, interagency, multi-institutional approach.

Cameron, M. F., Bengtson, J. L., Boveng, P. L., Jansen, J. K., Kelly, J. K., Dahle, S. P., Logerwell, E. A., Overland, J. E., Sabine, C. L., Waring, G. T., & Wilder, J. M. (2010). Status review of the bearded seal (*Erignathus barbatus*). U.S. Dep. Commer., NOAA Tech. Memo. NMFS-AFSC-211, 246 p.

Coffin, R., *et al.* (2010). First Trans-shelf-slope Climate Study in the U.S. Beaufort Sea Completed. *FITI Newsletter*.

Dlugokencky, E. J., & Bruhwiler, L. (2012). Sources of atmospheric carbon dioxide and methane in the Arctic. *Bulletin of American Meteorological Society*, 93(7), S144.

Grebmeier, J. M. (2012). Shifting patterns of life in the Pacific Arctic and Sub-Arctic Seas. *Annual Review of Marine Science*. Rev. Mar. Sci. 4(16.1-16.6).

Grebmeier, J. M., *et al.* (2010). Biological response to recent Pacific Arctic sea ice retreats. *EOS*, 91(18), 161-168.

Hezel, P.J., X. Zhang, C.M. Bitz, B. Kelly, and F. Massonnet (2012). Projected decline in snow depth on Arctic sea ice caused by progressively later autumn open ocean freeze-up this century, Geophysical Research Letters, VOL. 39, L17505, doi:10.1029/2012GL052794.

Huntington, H. P. (2009). A preliminary assessment of threats to arctic marine mammals and their conservation in the coming decades. *Mar. Policy* 33: 77-82.

Kelly, B. P. (2001). Climate change and ice breeding pinnipeds. In G. R. Walther, C. A. Burga, and P. J. Edwards (Eds.). *"Fingerprints" of Climate Change —Adapted Behavior and Shifting Species Ranges* (pp. 43-55). New York, NY, Kluwer Academic/ Plenum Publishers.

Kelly, B. P., Bengtson, J. L., Boveng, P. L., Cameron, M. F., Dahle, S. P., Jansen, J. K., Logerwell, E. A., Overland, J. E., Sabine, C. L., Waring, G. T., & Wilder, J. M. (2010). Status Review of the Ringed Seal (Phoca *hispida*). NOAA Tech. Memo. NMFS-AFSC-212.

Kovacs, K. M., Lyderson, C., *et al.* (2011). Impacts of changing sea ice conditions on Arctic marine mammals. *Marine Biodiversity*. 41, 181-194.

Metcalf, V., & Robards, M. (2008). Sustaining a healthy human-walrus relationship in a dynamic environment: challenge for comanagement. *Ecol App.* 18(2), S148-S156.

Moore, S. E., & George, J. C., *et al.* (2010). Bowhead whale distribution and feeding near Barrow, Alaska in late summer 2005-06. *Arctic* 63(2), 195-205.

Moore, S. E., & Huntington, H. P. (2008). Arctic marine mammals and climate change: impacts and resilience. *Ecol App.* 18(2), S157-S165.

NSIDC. http://nsidc.org/data/seaice_index/. Last accessed, Jan. 22. 2013.

Perovich, D. K., & Richter-Menge, J. (2009). Loss of sea ice in the Arctic. *Annual Review of Marine Science*, 1, 417-441.

Shakova, N., Semiletov, I., *et al.* (2010). Extensive Methane Venting to the Atmosphere from Sediments of the East Siberian Arctic Shelf. *Science*, 327, 1246-1250.

Stroeve, J., Holland, M. M., Meier, W., Scambos, T., & Serreze, M. (2007). Arctic sea ice decline: faster than forecast. *Geophysical Research Letters*, 34, L09501, doi:10.1029/2007GL029703.

3.2. Understand Terrestrial Ice Processes, Ecosystem Processes, Ecosystem Services, and Climate Feedbacks in the Arctic

Lead Author:
Shella Biallas, DOI
Agencies Partners:
DOE, DOI, NASA, NOAA, NSF, SI

Studying diminishing land ice, terrestrial ecosystem processes, ecosystem services, and climate feedbacks will contribute to a better understanding of the cumulative impacts of changes taking place in the Arctic and facilitate more informed decisions in the face of those changes.

Mass drainage of Arctic land ice from the Greenland ice sheet and glaciers and ice caps is a new and poorly understood problem with global implications (Milne *et al.* 2009). Climate feedbacks from ongoing ecosystem changes are only beginning to be understood but are expected to lead to further changes in the global climate system while affecting local communities'abilities to adapt to altered conditions on the ground.

The 2004 Arctic Climate Impact Assessment (ACIA 2005) outlined a number of high-priority science needs, including a better understanding of causes of increased Arctic land ice mass loss, changes likely to occur in terrestrial ecosystems, and how those changes will impact local communities. Currently, there are strong research partnerships among U.S. agencies operating in the Arctic. The Landscape Conservation Cooperatives (http://www.doi.gov/lcc/index.cfm), the North Slope Science Initiative (http://www.northslope.org/), and the Alaska Climate Science Center (http://www.doi.gov/csc/alaska/index.cfm), for example, are working to implement an applied-science program to inform management of Arctic natural resources.

The following five research areas will advance existing research participation, enhance coordinated approaches to understanding loss of land ice, climate feedbacks, terrestrial ecosystem processes and services, and support actions to address those changes.

3.2.1. Perform Glacial-Process Studies Targeting Specific Dynamic Regimes

Why Do This?

Mass loss from the Greenland ice sheet and Arctic glaciers and ice caps, as well as from the Antarctic Ice Sheet, has increased rapidly since the mid-1990s. The combined loss from polar land-ice now accounts for one-third to one-half of sea level rise (Cazenave and Llovel 2010; Church *et al.* 2011; Rignot *et al.* 2011; Jacobs *et al.* 2012). That loss is documented using space-based remote sensing—including altimetry, gravimetry, and Interferometric Synthetic Aperature Radar—as well as field and aircraft studies. Geodetic measurements of continental uplift and Earth rotation support these observations (e.g., Jiang *et al.* 2010; Nerem and Wahr 2011; Mitrovica and Wahr 2011).

The importance of research into the causes of increased Arctic land ice mass loss has been highlighted in numerous recent reports (e.g., http://www.acia.uaf.edu/, http://www.climatescience.gov/Library/ sap/sap1-2/final-report/default.htm, http://www. arctic.noaa.gov/reportcard). Loss of Arctic land-ice will result in potentially serious "downstream effects" including regional and global sea-level change (e.g., vulnerabilities of coastal populations, economic and national security infrastructure, and habitats). Understanding the causes of increased Arctic land ice mass loss, their connection to ocean and atmospheric variability, and the implications for current and future sea level will require process studies (this research area), observational studies (3.4), and modeling (3.5).

In the Arctic, the ice mass loss is due to increased surface melting and the acceleration, retreat, and thinning of tidewater glaciers in Greenland (van den Broeke *et al.* 2009) and Arctic tidewater glaciers (Arendt 2011). Though more research is needed to fully understand the complex drivers behind the observed acceleration of tidewater glaciers (Moon *et al.* 2012), evidence suggests that climate change is a major factor.

Evidence points to the marine margins of these glaciers as the region from which changes have propagated inland, but the drivers and mechanisms behind the acceleration are still unclear (Vieli and Nick 2011). The significance of this dynamic response has only recently been appreciated and is only beginning to be represented in current-generation ice sheet models. In the 2007 Intergovernmental Panel on Climate Change Fourth Assessment Report (IPCC AR4) report, this shortcoming was identified as the largest source of uncertainty in sea level change projections (Lemke *et al.* 2007).

Current projection estimates vary by more than an order of magnitude (Pfeffer *et al.* 2008; Price *et al.* 2011). In addition, the proximity of Greenland to the North Atlantic's dense water formation regions implies that an increasing discharge of freshwater from Greenland can potentially impact the large-scale overturning circulation of the North Atlantic (Dickson *et al.* 2008), a major player in the global oceanic heat transport, with far-reaching climatic implications.

The goals of the process studies are to identify and understand the relevant processes and develop/improve parameterizations for those processes that cannot be resolved in models. Six classe processes are identified, targeting Greenland outlet glaciers and Arctic tidewater glaciers:

1. Ice/ocean boundary layer and plume dynamics: Key measurements and modeling of the turbulent processes and their controls are needed to estimate submarine melt rates and develop appropriate melt rate parameterizations;
2. Fjord circulation and exchanges with the continental shelf: Integrated observational, modeling, and data analysis efforts are needed to understand how the fjord and shelf dynamics impact the ice/ocean boundary and the properties of sea ice, icebergs, and the ice mélange;
3. Glacial hydrology: Knowledge of the supraglacial, englacial, and subglacial hydrology, including discharge of freshwater into the fjord environment, is key to our understanding of ice flow, submarine melt rate, and plume dynamics;
4. Glacier dynamics: Process studies need to address the transition in ice flow from large catchment basins to narrow outlet or tidewater glaciers, in order to understand how the changes in stress-distribution and large-scale bed geometry influence the flow of ice and its supply to the terminus;
5. Calving: Calving plays a crucial role in both ice loss at the terminus and (indirectly) on the acceleration of inland ice flow, but its description remains elusive; and
6. Mass change.

Timeframe

Mid-term (3-5 years)

Expected Outcomes

- Improved understanding of basic principles and mechanisms controlling the identified physical processes.
- Direct observations of the meltwater plume at the ice/ocean interface, its spatio-temporal variability, accompanied by plume-resolving simulations.
- Observations of spatio-temporal variability of glacial hydrology (especially supraglacial lakes) and its relation to meteorological conditions, glacier flow, and meltwater plume variability.

- Developed and tested empirical relationship constrained by observations among submarine melt rate, glacier flow, fjord geometry, and circulation and ocean/atmosphere/sea-ice variability.
- Developed and tested empirical relationship constrained by observations between calving and environmental conditions as well as glacier internal stress balances.
- Based on observed relationships, formulate physically based parameterizations of unresolved processes for use in large-scale Earth system models (see 3.4).
- New methods and instrumentation systems, capable of monitoring subsurface melting, calving, subglacial discharge, and ice and sea-water properties in harsh environments.
- Enhance field campaigns with collection of paleo-proxies that provide evidence for past glacier variability.

Milestones

- Define observational requirements (essential variables, spatio-temporal sampling) for each of the components (ocean/fjord/ice-ocean interface/glacier/atmosphere) for process studies and sustained monitoring (NASA, NOAA; FY2014).
- Initiate 3-5 year interdisciplinary and interagency research initiatives that draw together observational, process modeling, and Earth system modeling expertise to accelerate progress on specific process parameterizations in U.S. Earth system models (NASA, NOAA, NSF; FY2014).
- Report strategies to feed results of process studies into sustained monitoring systems and Earth system models (DOE, NASA, NOAA, NSF; FY2014).
- Foster a U.S. and international research focus on ice/ocean interactions, especially with regards to the Greenland Ice Sheet. (NASA, NOAA, NSF; FY2015).

Science and Technology Gaps and Needs

- Development of instrumentation systems capable of functioning in harsh glacier terminus and fjord environments capable of withstanding the presence of sea-ice and iceberg, and observational techniques yielding robust measurements with high level of signal-to-noise ratio.
- Improvement of remote sensing technologies and underlying retrieval algorithms.
- Establishment of comprehensive, well-structured, and sophisticated databases and data formats to allow easy access and optimal use of data.
- Development of synergetic approaches (observational, theoretical, modeling, as well as multi-disciplinary) in order to pool resources and make progress in the directions outlined above.

3.2.2. Coordinate and Integrate Terrestrial Ecosystem Research Efforts

Why Do This?

Research on terrestrial ecosystems in the Arctic would greatly benefit from access to the range of existing data systems, information portals, and efforts to identify gaps in dissemination of research data and information. Resource and information sharing among agencies could be significantly enhanced by establishing formal agreements, such as the Memorandum of Understanding between NOAA and USGS (MoU 2011).

Existing sources of information and portals that provide a basis for comprehensive information sharing and coordination include:

- NSF Cooperative Arctic Data and Information Service (CADIS) and Advanced CADIS system (A-CADIS).
- Alaska Data Integration Working Group (ADIWG), formed to examine and address the technical barriers to efficiently integrate and share data within and among participating organizations.
- Polar Data Catalog maintained by the Canadian Cryospheric Information Network (CCIN).
- The North Slope Science Initiative/University of Alaska Geographic Information Network (NSSI/ GINA).
- USGS Science Portal.
- The Alaska Ocean Observing System (AOOS).
- The Arctic Council's Conservation of Arctic Flora and Fauna (CAFF) Circumpolar Biodiversity Monitoring Program (CBMP) is developing a distributed and decentralized geo-referenced, web-based portal for communications, data archiving, and information exchange.
- Arctic Portal established as part of the International Polar Year (IPY) activities and currently maintained by Iceland and various Arctic Council Working Groups.
- Arctic, Western Alaska, and Aleutians & Bering Sea Landscape Conservation Cooperatives (LCCs).
- Individual IARPC agencies (DOE, NASA, NPS[8], NWS[9], USGS) maintain substantial databases that inform cross-agency efforts.

Timeframe

Near-term (1-3 years)

Expected Outcomes

- Federal and Alaska state agency scientists identify the five most-pressing needs for scientific research in the terrestrial Arctic, in cooperation with Arctic Council efforts.
- Existing information sharing mechanisms, such as data.gov, AON/CADIS, or the NSSI/GINA data catalog, are fully used by scientists in planning and implementing terrestrial-ecosystem research in the Arctic to coordinate with other ongoing and planned efforts, and to expeditiously report their research findings.
- A single information delivery data hub is identified as the primary access point for various portals and databases of U.S. and international Arctic research activities to provide a point of access and collaboration for research activities.

Milestones

- Databases currently maintained by the Arctic LCC, NSF, NSSI, USGS, and non-governmental groups are made available on one website or via one access point (DOI, NSF; FY2013).
- Review completed and ongoing Arctic studies involving cross-disciplinary collaboration at the principle investigator level and identify any outstanding gaps and needs (NSF; FY2014).

Science and Technology Gaps and Needs

Federal agencies should agree upon and adopt a single access point to existing and future information-sharing databases and portals and share them with cooperating state, local, and international partners.

Stored long-term observing data and associated information should be accessible to and easily managed by contributing partners.

3.2.3. Identify and Study Key Sites Where Climate Feedbacks Are Active, Including Permafrost, Snow, Hydrates, and Glaciers

Why Do This?

The Arctic cryosphere regulates local, regional, and global climate and provides vital ecosystem services to communities. It also stores vast quantities of carbon (including in the form of methane) that may be released as temperatures increase and as ecosystem processes—such as wildland fires—change. A warming climate is having significant impacts on the Arctic cryosphere, such as decreased snow cover and thawing permafrost in the terrestrial environment. While terrestrial feedbacks in some locations have been studied, climate feedbacks in the region are generally still poorly understood (Arctic Monitoring and Assessment Program 2011).

The 2011 report, *Snow, Water, Ice, and Permafrost in the Arctic* (SWIPA), from the Arctic Monitoring and Assessment Program (AMAP) of the Arctic Council identified a number of areas where additional U.S.-supported research could improve understanding of the cryosphere (http://amap.no/swipa/). For example, better quantifying the intensity of cryosphere-climate feedbacks would lead to better predictions of the degree and rate of changes in the cryosphere and Arctic environment. It will also be critical to understand how Arctic communities will be affected by cryospheric changes, including those that influence migration patterns of subsistence species, utility of areas of traditional use, and the stability of infrastructure such as roads, airports, pipelines, and buildings. Other research is needed to understand how climate-driven changes in different parts of the cryosphere are affecting Arctic ecosystems—a critical first-step toward being able to predict how particular species (including those important for subsistence) will respond to new conditions.

Particular sites where climate feedbacks are observable can be identified by enhancing existing networks that build capacity for identifying, understanding, predicting, and responding to diverse environmental changes throughout the Arctic. For example, the Arctic LCC is identifying a suite of biophysical process topics most relevant to forecasting species and habitat response, which will guide recommendations for priority monitoring and modeling products. The Arctic LCC will recommend candidate sites for inclusion in a terrestrial ecosystem-monitoring network for northern Alaska that addresses the link-ages between physical drivers and biological responses. Implementing those recommendations will require the combined efforts and support of multiple agencies. The World Climate Research Program's Climate and Cryosphere (CliC) Project supports and coordinates international research on the interaction between the cryosphere and the rest of the climate system, focusing on identifying patterns and rates of change (http://www.climate-cryosphere.org/en/). Also, the International Network for Terrestrial Research and Monitoring in the Arctic (INTERACT) studies 32 sites in 14 countries, including all Arctic nations, with a goal of building capacity to improve understanding of environmental changes in the Arctic (http://www.eu-interact.org/).

Timeframe

Mid-term (3-5 years)

Expected Outcomes

- Cross-agency agreement on geographical areas in the U.S. Arctic where additional research on climate feedbacks and impacts on the cryosphere is most needed.
- Increased understanding of geophysical and ecosystem responses to changing climate and cryosphere change, which informs management decisions and subsistence uses.

Milestones

- Consult with local communities on the geographical areas of traditional use that are most impacted by changes in the terrestrial cryosphere, as well as the types of changes—permafrost thaw, coastal erosion, vegetation—most relevant to local communities (DOI; FY2013).
- Identify and compile Geographic Information Systems (GIS) data layers of existing climate feedback research in the Arctic and compare with areas of important traditional use (DOI; FY2012).
- Complete Wildlife Potential Habitat Forecasting Framework (WILDCAST) projections of potential land cover and habitat changes in Northern Alaska (USGS, FY2012).
- Assess existing tools and methods for measuring and mapping the effects of cryosphere changes on Arctic ecosystems and communities, including water levels, sea level rise, salt water intrusion, and coastal inundation (DOI, NSF; FY2013).
- Complete improved coastal map and high precision Digital Elevation Models (DEM) for western and northern Alaska in order to better understand coastal erosion, storm surges, and sea level rise. (DOI, NOAA, USCG; FY2013).

Science and Technology Gaps and Needs

Traditional knowledge can help detect changes in terrestrial Arctic ecosystems and guide adaptation. As the cryosphere changes, climate feedbacks become more complex, changes in Arctic systems will occur more rapidly, and integrating traditional knowledge into decisions will be as important as ever. Other needs include systematically collecting information about the effects of cryospheric change on human society and greater engagement between the scientific community and Arctic residents.

Improved networks are necessary to observe many of the long-term environmental changes taking place in the Arctic, including those pertaining to climate feedbacks (see section 3.4). Coordinating measures of changes in ecosystems and their components (e.g., species of concern) from *in-situ* and discrete observations with wider-ranging environmental parameters from remote instruments remains a challenge. Such coordination will improve the ability to relate geophysical change patterns to processes affecting ecosystems occurring across a range of spatial scales.

3.2.4. Investigate the Frequency and Severity of Wildland Fires in the Arctic

Why Do This?

Climate changes are affecting the frequency, extent, and severity of Arctic wildland fires and will have cascading effects on other ecosystem processes (Olsen *et al.* 2011). Wildfire is the primary disturbance in boreal forests of interior Alaska and is also a disturbance in the tundra regions of the U.S. Arctic. Fire frequency and severity are primary determinants of vegetative succession trajectories and, subsequently, the rates of carbon sequestration and loss in boreal ecosystems. Fire frequency and severity are likely to increase in tundra ecosystems as plant biomass and productivity increase with the lengthening growing season. Vegetation succession following fire in tundra ecosystems, however, is not well understood (Final Report of the Joint Fire Study Program 2011). In general, there is evidence that observed increase in tall shrub cover (Tape *et al.* 2006) is linked to an increase in fire frequency (Higuera *et al.* 2008). Fire regimes and impacts have been observed on fine spatial scales at point locations in Alaska, but the collective impacts of wildland fire on vegetation, carbon, wildlife, air quality, permafrost degradation, and biogeochemical cycles across the landscape are poorly understood, yet critical to regional fire management strategies.

The 2007 Anaktuvuk River fire was the largest tundra fire to take place in the United States since recording began in 1950 and had major implications both for local ecosystems and the global carbon system. Several agencies have funded investigations of the severity of and ecological response to the Anaktuvuk River fire. That type of research, however, has not been a priority for resource managers. A recent study found the Anaktuvuk fire released carbon into the atmosphere about 100 times faster than it usually escapes from the ground in the Arctic summer, and released more than two million tons of CO_2 (Mack 2011). The fire was deep and severe for a tundra burn (Mack *et al.* 2011), and lichen cover was drastically reduced, thereby reducing winter forage for caribou. Because caribou foraging on winter range tend to avoid burned areas for up to 50 years (Joly 2007), large-scale burns may change seasonal distribution of the animals, thereby changing their availability to subsistence users.

Timeframe

Mid-term (3-5 years)

Expected Outcomes

- An integrated understanding of the trends and impacts of Arctic wildland fires to inform policy and land-management decisions.

Milestones

- Identify and inventory existing scientific research on Arctic wildland fires (DOI; FY2012).
- Consult with local communities and indigenous groups on science needs pertaining to Arctic wildland fires and their impacts on cultural and subsistence needs (DOI; FY2012).
- Develop strategies/projects to identify succession stages of tundra communities following a wildfire (DOI, NSF; FY2013).

- Ensure coordination in the development of models that incorporate feedback from fire models with models of surface vegetation and organic layer properties and permafrost and soil conditions, incorporating hydrologic information as appropriate (DOE, DOI, NSF; FY2016).

Science and Technology Gaps and Needs

While the 2007 Anaktuvuk River fire was an event of unprecedented scale on the North Slope, it is not clear whether ongoing climate changes will lead to additional severe fires. Paleoecological work assessing fire-deposited charcoal in lake cores is underway and should contribute to improved under-standing of fire regime in that area. As fire-return intervals have historically been several hundreds of years, detecting less than very dramatic changes in long-term fire regime, or fire severity, may not be possible. Credible prediction of the potential effects of changes in fire regime, or fire management policy, on caribou distribution and subsistence use will likewise be challenging. The Alaska Integrated Ecosystem Modeling project, through the Alaska Climate Science Center and the Western, Arctic, and Northwest Interior Forest LCCs, is creating a framework to link existing models to produce a single integrated platform that simulates vegetation succession, disturbance regimes, hydrology, and permafrost dynamics. With broad support, this work can contribute to state-wide maps that forecast ecological conditions under specified climate scenarios. Data also are needed to improve response to fires and understanding of fire behavior include greater coverage in lightning-strike detection, particularly in the central Arctic, and enhanced distribution of weather-station observation sites that collect temperature and precipitation data.

3.2.5. Conduct Socio-Economic Research to Understand Ecosystem Services As Increased Warming Changes the Arctic Tundra

Why Do This?

More than 86% of communities in rural Alaska have been affected by serious erosion and flooding (GAO 2003). Forced relocation of some communities comes at a high cost, and sites must be selected that will remain stable indefinitely. Local residents are being forced to address immediate needs, develop adaptation strategies for the future, and also balance social, cultural, and economic demands and expectations simultaneously (see section 3.6).

Additional environmental stressors brought on by increased human activity in the area are also affecting Arctic communities. Such rapid change affects supply of subsistence foods and storage capability, integrity of local infrastructure, and social and cultural systems (http://www.arctichealth.org /ccNorthernCommunities). At the same time, economic and political pressures to fulfill global energy needs place a burden on Arctic residents.

More research is needed on the social and economic impacts of ongoing environmental changes and their implications for local communities. Such information will provide the framework and data needed to develop proactive plans for protecting, managing, and adapting to a fragile and changing Arctic environment.

Timeframe

Mid-term (3-5 years)

Expected Outcomes

- A survey of the state of knowledge on terrestrial ecosystem services in the Arctic including identification of gaps in research on the importance of ecosystem services for Northern communities.
- A set of robust and reliable social indicators are established, and relevant data are collected to predict, monitor, and mitigate the effects of climate change on Alaskan Arctic communities.

Milestones

- Support the outcomes and recommendations of the Arctic Social Indicators Project (http:// www.svs.is/asi/Implementation/Project %20description%20II.htm) within the United States.
  - Develop a meta-database, published electronically, that identifies baseline Arctic social indicators already monitored by national agencies (DOI, NSF, SI; FY2013).
  - Establish an international task force of Arctic researchers to ensure social data are collected in a way that allows statistically valid comparisons among Arctic communities (DOI, DOS, NSF, SI; FY2012).
  - Collect indigenous and non-indigenous regional data on the socio-economic implications of climate change at 5-year intervals (DOI, NIH, NSF, SI; FY2013).

Science and Technology Gaps and Needs

Socio-economic data at the village and regional level are lacking as are direct investigations of the links between cause and social effects. The mechanisms of cause and effect are often complex and multifactoral. Small population size at the village and region level may preclude acquisition of statistically significant data for some indices.

References

ACIA (2005). Arctic Climate Impact Assessment: Scientific Report. 1042 pp. Cambridge University Press, UK.

Arctic Health. University of Alaska—Anchorage. Compilation of Alaska Northern Communities Resilience and Adaptation Research—http://www.arctichealth.org /ccNorthernCommunities.

Arctic Monitoring and Assessment Program. (2007). Oil and Gas in the Arctic: Effects and Potential Effects Volume 1 (Chapter 2—Social and Economic Effects).

Arctic Monitoring and Assessment Program. (2011). Working chapter 11.1 of Arctic Council SWIPA report.

Arendt, A. A. (2011). Assessing the status of Alaska's glaciers. *Science*, 332, 1044-1045. doi: 10.1126/ science.1204400.

Cazenave, A., & Llovel, W. (2010). Contemporary sea level rise. *Annu. Rev. Mar. Sci.*, 2, 145-173. doi:10.1146/ annurev-marine-120308-081105.

Church, J. A., *et al.* (2011). Revisiting Earth's sea-level and energy budgets from 1961 to 2008. *Geophys Res Lett.*, 38, L18601.

Dickson, R. R., Meincke, J., & Rhines, P. (2008). Arctic-Subarctic Ocean Fluxes. Springer.

Government Accountability Office Report. (2003). ALASKA NATIVE VILLAGES: *Most Are Affected by Flooding and Erosion, but Few Qualify for Federal Assistance*. http://www.gao.gov/new.items/ d04142.pdf. Last accessed Jan. 22, 2013.

Higuera, P.E., *et al.* (2008). Frequent Fires in Ancient Shrub Tundra: Implications of Paleorecords for Arctic Environmental Change. *PLoS ONE* 3(3), e0001744. doi:10.1371/journal.pone.0001744.

Jacob, T., Wahr, J., Pfeffer, W. T., & Swenson, S. (2012). Recent contributions of glaciers and ice caps to sea level rise. *Nature*, 482, 514-518. doi:10.1038/nature10847.

Jiang, Y., Dixon, T.H., & Wdowinski, S. (2010). Accelerating uplift in the North Atlantic region as an indicator of ice loss. *Nature Geosci.*, 3, 404-407. doi:10.1038/ngeo845.

Joly, K., Bente, P., & Dau, J. (2006). Response of Overwintering Caribou to Burned Habitat in Northwest Alaska. *Arctic*. 60(4) (December 2007) 401–410.

Lemke, P., Ren, J., Alley, R. B., Allison, I., Carrasco, J., Flato, G., Fujii, Y., Kaser, G., Mote, P., Thomas, R.H., & Zhang, T. (2007). Observations: Changes in Snow, Ice and Frozen Ground. In: Climate Change 2007: The Physical Science Basis. Contribution of Working Group I to the Fourth Assessment Report of the Intergovernmental Panel on Climate Change. [Solomon, S., D. Qin, M. Manning, Z. Chen, M. Marquis, K.B. Averyt, M. Tignor and H.L. Miller (eds.)]. Cambridge, UK. & New York, NY: Cambridge University Press.

Mack, M. C., Bret-Harte, M. S., Hollingsworth, T. N., Jandt, R. R., Schuur, E. A. G., Shaver, G. R., & Verbyla, D. L. (2011). Carbon loss from an unprecedented Arctic tundra wildfire. *Nature* 475, 489-492.

Memorandum of Understanding between the USGS and NOAA to Coordinate and Cooperate in Activities Involving Physical and Biological Sciences and Environmental Studies. Signed Jan. 9, 2011.

Milne, G. A., Gehrels, W. R., Hughes, W., & Tamisiea, M. E. (2009). Identifying the causes of sea-level change. *Nature Geosci.*, 2, 471-478.

Mitrovica, J. X., & Wahr, J. (2011). *Ice Age Earth Rotation.* Annu. Rev. Earth Planet. Sci., 39, 577–616. doi:10.1146/annurev-earth-040610-133404.

Mix, A. C., Samelson, R., & Padman, L. (Eds.). (2012). Interdisciplinary Approaches to Understanding Atmosphere/Ocean/Ice Shelf/Ice Sheet Interactions. A Workshop Report to NSF.

Moon, T., Joughin, I., Smith, B. & Howat, I. (2012). 21st-Century evolution of Greenland outlet glacier velocities. *Science*, 336, 576-578.

Nerem, R. S., & Wahr, J. (2011). Recent changes in the Earth's oblateness driven by Greenland and Antarctic ice mass loss. *Geophys. Res. Lett.*, 38, L13501. doi:10.1029/2011GL047879.

Olson, D., *et al.* (2011). Final Report to the Joint Fire Science Program: Compiling, Synthesizing and Analyzing Existing Boreal Forest Fire History Data in Alaska. July 2011. http://www.firescience.gov/ projects/06-3-1-26/project/06-3-1-26_final_report.pdf. Last accessed Jan. 22, 2013.

Pfeffer, W. T., Harper, J. T., & O'Neel, S. (2008). Kinematic constraints on glacier contributions to 21st-century sea-level rise. *Science*, 321, 1340-1343.

Price, S. F., Payne, A. J., Howat, I. M., & Smith, B. E. (2011). Committed sea-level rise for the next century from Greenland ice sheet dynamics during the past decade. *P. Natl. Acad. Sci. USA*, 108, 8978–8983.

Rignot, E., Velicogna, I., van den Broeke, M. R., Monaghan, A., & Lenaerts, J. (2011). Acceleration of the contribution of the Greenland and Antarctic ice sheets to sea level rise. *Geophys. Res. Lett.*, 38, L05503. doi:10.1029/2011GL046583.

Tape, K., Sturm, M., & Racine, C. (2006). The evidence for shrub expansion in Northern Alaska and the Pan-Arctic. *Global Change Biology*, 12, 686-702.

The Arctic Social Indicators Project. http://www.svs.is/asi/Implementation/Project%20description%20II.htm van den Broeke, M., *et al.* (2009). Partitioning recent Greenland mass loss. *Science*, 326, 984. doi: 10.1126/science.1178176.

Vieli, A., & Nick, F. M. (2011). Understanding and modeling rapid dynamical changes of tidewater outlet glaciers: issues and implications. *Surv. Geophys.* 32(4-5), 437–458. doi:10.1007/s10712-011-9132-4.

3.3. Understand Atmospheric Surface Heat, Energy, and Mass Balances

Lead Authors:
Richard Eckman, NASA
Taneil Uttal, NOAA
Sandy Starkweather, NOAA
Agency Partners:
DOE, NASA, NOAA, NSF

Year-to-year and longer-scale trends and variability in surface-air temperature tend to be larger in the Arctic than in other parts of the globe. This Arctic amplification phenomenon is recognized as an inherent characteristic of the global climate system (Serreze and Barry 2011); the causes are believed to include complex interactions associated with heat exchange between the atmosphere and ocean (with its changing sea-ice extent), meridional heat transport, and radiative forcing from atmospheric constituents. Arctic atmospheric processes are influenced by unique features (polar night, high albedo surfaces, and atmospheric stability) that can change the sign and magnitude of aerosol and cloud radiative forcing relative to low latitudes. Important uncertainties in the sign and magnitude of these forcings (e.g., IPCC AR4) provide the rationale for coordinating and improving our integrated understanding of Arctic atmospheric processes. This plan addresses those uncertainties through milestones aimed at improving process representations in models, reducing uncertainty in model outputs, and developing long-term observational data sets.

Remote sensing and *in-situ* observations at long-term observatories complement each other and contribute to documenting and understanding long-term trends. NASA and NOAA remote sensing assets offer an unique resource for observing Arctic atmospheric composition and radiative forcing. DOE, NSF, and NOAA support key ground-based observatories providing long-term data sets and campaign studies that are critical resources for addressing key climate model uncertainties such as clouds, aerosols, and impacts of short-lived climate forcers (SLCFs). These land-based observatories are also valuable resources for validating satellite observations. Data from these observatories also are available through the International Arctic System for Observing the Atmosphere (IASOA). During the International Polar Year, IASOA developed a major new international observatory facility in Tiksi, Russia, that significantly extends circumpolar coverage. DOE, NASA, NSF, and NOAA field and aircraft campaigns have contributed to an integrated understanding of the Arctic atmosphere and improved model parameterizations and the value of remote-sensing data products. DOE- and NSF-supported predictive modeling contributes to long-term understanding of regional and global sensitivities to aerosol loading and cloud processes, have revealed the relative contributions of short-lived climate forcers (discussed in more detail in the next section), and have helped to develop an integrated picture of the atmospheric interactions with terrestrial, oceanic, and cryospheric systems. Coordinated approaches around the Arctic, particularly those supported by international partnerships, are critical to developing a regionally coherent understanding of both how and why the Arctic atmosphere is changing. Jointly, these activities are improving our understanding of the unique role of the Arctic atmosphere in influencing the Arctic surface system as well as regional and global weather and climate.

3.3.1. Improve Understanding of Short-Lived Climate Forcers (SLCFs); Source Regions, Direct and Indirect Effects, and Net Impact on Arctic Warming

Why Do This?

Short-lived climate forcers (SLCFs) including black carbon, methane, and ozone are atmospheric constituents with relatively short residence times (days to years) and that warm the climate. These forcers are thought to have an enhanced influence on Arctic radiative forcing relative to mid-latitudes (Quinn *et al.* 2008). For example, black carbon can change Arctic radiation balances by either direct forcing by the radiative profile of the atmosphere, by aerosol-cloud indirect effects, or by lowering the albedo of (typically) bright Arctic surfaces after deposition or, potentially, by hastening the thaw of snow and ice. Investigating the relative contributions of these three mechanisms and the effects of black carbon requires additional attention. The Arctic contains vast amounts of methane locked in permafrost deposits and marine hydrates, with an uncertain potential for release into the atmosphere. As a greenhouse gas, methane is approximately twenty times more effective at trapping heat than is carbon dioxide over a 100-year period. Understanding current methane emissions and potential scenarios under a warmer Arctic is imperative. Many global circulation models do not take into account carbon feedback loops from Arctic tundra, where warming causes carbon release from thawing and decaying tundra that, in turn, could further accelerate carbon release. Ozone is both an air pollutant that impacts human health as well as a greenhouse gas. Both Arctic and remote sources produce ozone and its precursors. Boreal forest fires and increasing human activity will increase ozone precursors in the region. Further research is required to understand sources of ozone precursors as well as the oxidation capacity of the Arctic atmosphere. Ozone depletion events have been related to the deposition of mercury, which is a significant and toxic pollutant.

Existing agency capabilities to study SLCF's include satellite instruments that monitor the long-range transport of mid-latitude pollution to the Arctic, in addition to detecting fires and their smoke plumes. These instruments measure aerosol optical depths, other aerosol properties, and collocated cloud properties. The Moderate Resolution Imaging Spectroradiometer (MODIS) and the Multi-angle Imaging SpectroRadiometer (MISR) provide an aerosol record dating from 2000, while the Atmospheric Infrared Sensor (AIRS) provides a carbon monoxide record from 2002. The Cloud-Aerosol LIDAR and Infrared Pathfinder Satellite Observation (CALIPSO) satellite Cloud-Aerosol LIDAR with Orthogonal Polarization (CALIOP) instrument provides high-resolution vertical profiles of aerosols and clouds from 2006. Flying in formation with Aqua (MODIS, AIRS), Aura and GCOM-W1, the multi-sensor "A-Train" provides near simultaneous measurements of a variety of parameters. IASOA network observatories include proxy and direct measurements of black carbon, as well as direct measurements of methane and ozone at limited locations. About a dozen Aerosol Robotic Network (AERONET) and AeroCan (AERosol CANada) sunphotometer sites located at latitudes poleward of 60°N monitor aerosol amount and type in seasons when the sun is above the horizon. The DOE Next Generation Ecosystem Experiment (NGEE) is focusing on measurements for improving the model simulations of climate-related emissions of SLCFs. DOE, NASA, and NOAA aircraft campaigns have also contributed to an integrated understanding of the Arctic atmosphere.

The NASA Arctic Research on the Composition of the Troposphere from Aircraft and Satellites (ARCTAS), for example, provided new characterization of bidirectional reflectance distribution functions for aerosols over Arctic surfaces, validated CALIOP LIDAR aerosol sensing for a range of conditions, and provided detailed characterization of the optical properties of aerosols from boreal fires.

Timeframe

Mid-term (3-5 years)

Expected Outcomes

- Sustained and improved satellite-observation capabilities focused on SLCF's (including black carbon, methane, and ozone).
- Enhanced *in-situ*, long-term observations of SLCF's including a methane-observing network.
- Improved modeling of SLCF transport and lifetime.

Milestones

- Support process studies and campaigns to validate current satellite measurements, such as aerosol products from MODIS, MISR, and Visible Infrared Imager Radiometer Suite (VIIRS) and methane and near-surface ozone from the Tropospheric Emission Spectrometer (TES) on Aura (DOE, NASA, NOAA; FY2016).
- Develop pan-Arctic synthesis of SLCF's from current observations focused on concentrations, sources, and radiative impacts (NOAA, NSF; FY2014).
- Develop needs assessment for improved transport-modeling capability (DOE, NSF; FY2014).
- Support SLCF's source identification through transport and regional modeling using satellite and suborbital data to constrain the models (DOE, NSF; FY2017).
- Support black carbon source identification through chemical composition measurement at key observatory locations and aerosol mapping from space with MODIS and MISR (DOE, NASA, NOAA; FY2017).
- Support black carbon radiative impact studies through *in-situ* measurements at key observatory locations and modeling of light scattering, absorption, and aerosol optical depth (DOE, NOAA; FY2017).
- Develop needs assessment for an Arctic methane-observation network (DOE, NASA, NSF; FY2013).

- Increase spatial density of Arctic methane measurements (DOE, NOAA, NSF; FY2017).

Science and Technology Gaps and Needs

NASA operates a variety of sensors that monitor trends and transport of SLCF's. Satellite retrievals of SLCF's offer poor resolution near the Earth's surface and must be integrated with higher-resolution *in-situ* measurements. Agencies should support field validation campaigns to obtain direct information about aerosol and surface properties as well as other activities that reduce uncertainty in retrievals by satellites.

Large spatial gaps exist in the current network of *in-situ* measurements for SLCF's. Ground-based observatories could make stronger contributions to understanding the role of black carbon by building capacity to monitor the mass concentrations and chemical composition of aerosols in addition to current light-absorbing proxy measurements. Locations should be based on sampling representative air-mass trajectories. Given the spatial extent and diversity of methane sources, a much broader network of sustained methane measurements is required to monitor long-term methane trends, understand the processes that emit methane, and better constrain inverse model studies of the methane budget.

3.3.2. Improve Understanding of Processes That Control the Formation, Longevity, and Physical Properties of Arctic Clouds, Including the Effects of—and Sensitivities to—Aerosols

Why Do This?

Unlike other atmospheric features, clouds and aerosols are unevenly distributed in space and participate in highly integrated processes. Clouds, particularly those at low levels, occur frequently throughout the Arctic. They are particularly susceptible to aerosol influences on both liquid-droplet and ice-crystal nucleation, which impacts cloud formation, persistence, physical properties, and precipitation. Substantial uncertainty surrounds which modes of nucleation are operating under the varying conditions in the Arctic and how they are linked to aerosol composition and sources. Due to the prevalence of low sun angles and high surface albedos, Arctic aerosol and cloud radiative forcings typically have different signs than at lower latitudes. Specifically, they tend to warm the surface predominantly by trapping long-wave radiation more efficiently than they cool the surface by reflecting sunlight. Aerosols can change cloud cover, thickness, and brightness. These perturbations, thus, have direct implications for the net radiative balance at the Earth's surface and top of the atmosphere. Arctic long-wave radiative forcing from low-lying clouds has been identified as an important controller of onset and duration of surface melt impacting the mass budgets of sea ice and ice sheets and the seasonal extent of snow cover. Cloud precipitation processes are a fundamental component of the Arctic hydrological cycle. Due to numerous, poorly known, complex processes involving clouds and aerosols, their misrepresenstation in models lead to large uncertainties in climate simulations of (i.e., IPCC AR4). To improve these modeling deficiencies, it is imperative to characterize and understand basic cloud and aerosol properties and their interactions within the system.

U.S. agencies (DOE, NASA, NOAA, and NSF) employ satellite, ground-based, and *in-situ* assets to observe Arctic clouds and aerosols. Passive measurements from space are challenging, particularly during winter, and over snow and ice surfaces, due to low optical and thermal contrast among clouds, aerosols, and the underlying surface. Using satellites to assess the impact of aerosols on Arctic clouds is an even greater challenge for passive remote-sensing instruments due to persistent cloud cover, poor vertical discrimination, bright surfaces, relatively low aerosol column abundances, and low solar-illumination angles, but these measurement limitations are significantly improved by combining passive instruments, such as MODIS (measuring ambient radiation), with active instruments, such as CALIOP and CloudSat (measuring emitted LIDAR and radar returns). In addition to providing an unparalleled aerosol record going back to 2000, the MODIS and MISR satellite sensors monitor the long-range transport of pollution and smoke into the Arctic. Ground-based observations, however, provide a very important dataset of Arctic clouds. These measurements are being used to improve our knowledge of cloud processes, and thereby improve our ability to represent these processes in numerical models. Furthermore, these observations are also providing detailed climatologies, albeit in only a small number of locations, of cloud occurrence and phase as a function of height, as well as cloud particle size and water path.

Important progress has also been made towards understanding cloud-aerosol interactions and properties via aircraft campaigns such as the recent NASA Arctic Research on the Composition of the Troposphere from Aircraft and Satellites (ARCTAS), the DOE Indirect and Semi-Direct Aerosol Campaign (ISDAC), and the NOAA Aerosol, Radiation, and Cloud Processes affecting Arctic Climate (ARCPAC). For example, ARCTAS provided new characterizations useful for energy balance assessment and a better understanding of aerosol radiative effects, which help to improve satellite measurements. ARCPAC and ISDAC both provided important new perspectives on aerosol composition and transport sources with important implications for cloud processes.

The International Arctic System for Observing the Atmosphere (IASOA) is a pan-Arctic consortium of flagship ground-based observatories that are supported by DOE, NOAA, NSF, Environment Canada (EC), the Russian Federal Service for Environmental and Hydrometeorolgical Monitoring (*Roshydromet*), the Finnish Meteorological Institute (FMI), and other government and non-government contributors from Arctic and non-Arctic countries. The observatories operate sophisticated instruments that support sustained, high-resolution, and simultaneous observations of clouds, aerosols, atmospheric structure, and the surface-energy balance. These year-round observatories provide excellent platforms for both long-term and campaign-based process studies. DOE's Atmospheric Radiation Measurement (ARM) Climate Research Facility in Barrow, Alaska, for example, provides a long-term record of highly valuable simultaneous measurements of cloud microphysical and macrophysical parameters, aerosols, and surface radiation. Recent additions of scanning radars, powerful LIDARs, flux measurements, and aerosol instruments further enhance ARM's measurement capabilities over an extended volume near Barrow. Measurements from that site have been the basis for several successful international model inter-comparison projects focused on cloud-resolving and single-column model simulations, whose goals are to improve Arctic cloud microphysics parameterizations and to better understand indirect effects of aerosols. Instrumentation at the new ARM site at Oliktok, Alaska, mirrors the Barrow site with the added capability of instrumented unmanned aerial vehicles (UAVs). The focus of work at the Oliktok site is to examine clouds and aerosols over land, sea, and ice, as well as the coupled atmosphere-cloudterrestrial Arctic systems. The NSF has established an ARM-like site at Summit Station on the Greenland Ice Sheet. NOAA and EC operate a similar

ARM-like site at Eureka, Canada. Ground-based observatories, in general, contribute unique and valuable information for use in model evaluation and development, for validation of satellite observational methods, and for long-term monitoring of Arctic atmospheric properties.

Timeframe

Mid-term (3-5 years)

Expected Outcomes

- Evaluation, improvement, and development of cloud and aerosol parameterizations in climate and weather prediction models.
- Sustained and improved satellite-observation capabilities of cloud and aerosol properties, with verification using surface observations.
- Sustained and enhanced ground-based observations including simultaneous measurements of clouds and aerosols.
- Improved understanding of how clouds respond to changing levels of aerosols and sea ice.
- Synthesis of data sets that provide detailed descriptions of clouds and aerosols.

Milestones

- Support sustained and enhanced ground-based measurements of cloud and aerosol properties (DOE, NASA, NOAA, NSF; FY2017).
- Conduct intensive, short-term, ground-based, and airborne field experiments to quantify the impact of aerosols on clouds, conduct detailed process studies, and provide validation data sets for remote sensing data (DOE, NASA, NOAA, NSF; FY2017).
- Support synthesis activities to develop long-term observational cloud and aerosol data sets from ground-based and satellite platforms to evaluate model parameterizations (DOE, NASA, NOAA, NSF; FY2017).
- Use observational data sets to constrain process-model studies and conduct detailed model inter-comparisons to advance parameterization development (DOE, NSF; FY2017).
- Support laboratory studies to examine cloud-particle nucleation processes (DOE, NSF; FY2014).
- Support observations and modeling activities to improve understanding of transport of aerosols from remote regions to the Arctic (NASA, NSF; FY2017).

Science and Technology Gaps and Needs

At many scales, models do not represent cloud and aerosol processes well. This general lack is based upon observational data sets of insufficient length, complexity, and spatial representation, as well as insufficient process-resolving models. Specific gaps in understanding include processes related to aerosol sources and transport, basic cloud-aerosol interactions, cloud-phase partitioning, the influence of heat and moisture advection on cloud formation, interactions between clouds and atmospheric structure, and determining the relative contributions of different cloud types to precipitation totals.

To address the need to advance physical parameterizations in numerical climate and weather-prediction models, measurements related to cloud-aerosol-radiation processes at DOE, IASOA, NASA, NOAA, and NSF observatories should be sustained and enhanced. These measurements should be supplemented by intensive campaigns (DOE, NASA, NOAA, and NSF) focused on enhancing ongoing, long-term measurements and targeting specific processes or hypotheses. To expand the spatial footprint of these observations, DOE, NASA, NOAA, and NSF are funding technology development through the Small Business Innovation Research (SBIR) program to enable cloud and aerosol measurements from UAVs and balloon-borne platforms. Enhanced emphasis should also be placed on validation and refinement of satellite measurement technologies for observing aerosol characteristics and transport, as well as characterizing clouds throughout the troposphere (NASA, NOAA). Satellite research would also greatly benefit from passive measurements from the Suomi National Polar-orbiting Partnership (NPP) mission launched in 2011, the forthcoming Joint Polar Satellite System-1 (JPSS-1), and international collaboration with European countries and Japan around future active measurements from the EarthCare mission. Ideally, development of a more advanced satellite-based, multi-angle, multi-spectral, polarimetric imager should be considered. Finally, where appropriate, agencies should support evaluation, utilization, and advancement of smaller-scale models that can be used in both process studies and as intermediaries for parameterization development.

3.3.3. Develop an Integrated Understanding of Arctic Atmospheric Processes, Their Impact on the Surface Energy Budget, and Their Linkages with Oceanic, Terrestrial, and Cryospheric Systems

Why Do This?

Numerical models are used to understand and predict important processes such as the decline of Arctic sea ice, linkages between Arctic conditions and lower-latitude weather, and the general amplification of climate change in the Arctic. Developing models of sufficient quality, however, relies on building a system-level understanding of Arctic climate that includes detailed knowledge about the Arctic atmosphere and surface and their many interacting processes. Current models face considerable difficulties when representing Arctic atmospheric processes related to boundary layer structure, cloud formation, and aerosol-cloud interactions—all of which interact critically with the surface. The general inabilities of models to accurately represent the observed decline in Arctic sea ice and to properly capture its causes represent one tangible example. Such modeling difficulties result directly from limited, system-level observations and could be remedied by comprehensive and coordinated measurements of all contributing components.

Process- and system-level understanding requires coordination and expansion of observational capabilities along with complementary process-modeling activities. Satellite measurements—a cornerstone of observation —provide valuable spatial coverage of many key parameters with linkages to those from lower-latitudes. They are often inadequate, however, for quantifying basic parameters such as surface temperatures and providing the level of detail needed for coordinated radiation, cloud, aerosol, and other atmospheric measurements. Additionally, many polar-orbiting satellites take insufficient measurements in the area within 800 kilometers of the North Pole. Although ground-based observations and aircraft campaigns may be spatially and/or temporally limited, they provide the types of measurements needed to characterize many of the necessary processes in high detail. Most existing knowledge on Arctic atmosphere-surface interactions is biased towards land-based or coastal observations and processes. Thus, the challenge is to coordinate existing and new interagency observational abilities from satellite, aircraft, and the ground to produce the comprehensive, process-level observations of the Arctic system needed to improve numerical models.

Timeframe

Mid-term (3-5 years)

Expected Outcomes

- Sustained and improved satellite-observation capabilities focused on atmospheric interactions with the surface and coordination of satellite-measurement capabilities.
- Improved understanding of the two-way relationship between sea ice and clouds.
- Sustained and enhanced ground-based observations emphasizing simultaneous measurements of clouds, aerosols, atmospheric structure, and surface-energy budget in land- and ocean-based environments.
- Improved representation of Arctic systems in climate and weather-prediction models.

Milestones

- Support model-component development and advancement of fundamental knowledge of the key processes that regulate aerosol and cloud impacts on the atmospheric- and surface-energy budgets (DOE, NASA, NOAA, NSF; FY2015).
- Support research activities that integrate Arctic processes in regional and global models (DOE, NASA, NOAA, NSF; FY2017).
- Coordinate interdisciplinary campaigns to study the Arctic climate system as a whole (DOE, NSF; FY2017).
- Increase use of UAV (unmanned aerial vehicles) platforms for targeted observations of Arctic processes (DOE, NASA, NOAA; FY2015).

Science and Technology Gaps and Needs

Three prominent gaps exist in research on Arctic atmospheric processes and their interactions with the broader Arctic climate system: 1) a lack of comprehensive observations at specific locations; 2) a general dearth of process-level observations over sea-ice environments; and 3) limited routine observations in the central Arctic Basin.

Barrow, Alaska, is a primary hub for Arctic research in the United States. Benefiting from substantial observatory and campaign efforts by DOE, NASA, NOAA, and NSF, Barrow likely contains the most comprehensive Arctic atmospheric observatory capabilities worldwide and, in many ways, is a model for such observations. This comprehensive level, however, is not currently available at other key Arctic IASOA observatories operated in Canada, Greenland, Russia, and Northern Europe. Efforts should be made to broaden observational capabilities at additional Arctic locations, leveraging existing industrial monitoring sites where appropriate. The new ARM site at Oliktok, Alaska, with instrumentation mirroring the Barrow site—but with the added capability of instrumented UAVs—should attract other agency contributors. Additional instrumentation from other agencies, such as NASA, NOAA, and NSF, would provide more-comprehensive observations of clouds and aerosols over land, sea, and ice, as well as the coupled atmosphere-cloud-terrestrial Arctic systems.

The central Arctic Basin represents a substantial spatial coverage gap that limits models of important processes, such as those related to sea-ice decline. Multi-year, detailed, and comprehensive measurements, extending from the ocean through the sea ice and into the atmosphere, are critically needed to provide a process-level understanding of the complex regional systems of interactions and feedbacks that cannot be gained by land- or space-based observations. One option is to establish an intensive, multi-year, surface-based observatory in the central Arctic Basin via an icebreaker-supported ice sta-tion. Such an ambitions endeavor will require the coordinated support of multiple U.S. agencies and international partners. The 1997-1998 Surface Heat Budget of the Arctic Ocean campaign (SHEBA), with participation by DOE, NASA, NOAA, and NSF was a good example of interagency coordination on a ship-based campaign that provided a first look into many central Arctic processes. Experience from SHEBA and other campaigns over the Arctic sea ice has shown that the most valuable data have been obtained from campaigns making comprehensive, multi-disciplinary measurements during one or more annual cycles; now in the "new Arctic" is an opportune time to look a second look at a substantially changed central Arctic.

Finally, operational models and model re-analyses assimilate observational data to improve model accuracy. The central Arctic Basin is, however, sparsely populated with routine observational inputs for these models. First, efforts should be made to improve operational satellite products through ground and aircraft-based validation and algorithm development to provide improved constraints on model performance (NASA and NOAA). Second, expanded operational observations of basic Arctic atmospheric properties are needed, such as those provided by additional radiosonde stations or via routine dropsonde observations. UAVs offer great promise for cost-effective means to make sustained measurements over the central Arctic (DOE, NASA, and NOAA). These observational enhancements have the potential to reduce uncertainty in operational models and re-analysis products over both the Arctic and lower latitudes.

References

Darby, L. S., *et al.* (2011). International Arctic System for Observing the Atmosphere (IASOA). In I. Krupnik, *et al.* (Eds.), *Understanding earth's polar challengers: International Polar Year 2007-2008*, Edmonton, Alberta, Canada: University of the Arctic.

Jacob, D. J., *et al.* (2010). The Arctic Research of the Composition of the Troposphere from Aircraft and Satellites (ARCTAS) mission: design, execution, and first results, *Atmos. Chem. Phys.*, 10, 5191-5292.

McFarquhar, G. M., *et al.* (2011). Indirect and Semi-Indirect Aerosol Campaign: The Impact of Arctic Aerosols on Clouds, *Bull. Amer. Meteor. Soc.*, 92, 183-201.

Solomon, S., Qin, D., Manning, M., Chen, Z., Marquis, M., Averyt, K. B., Tignor, M., and Miller, H. L. (Eds.). (2007). Contribution of Working Group I to the Fourth Assessment Report of the Intergovernmental Panel on Climate Change. Cambridge, United Kingdom and New York, NY, USA Cambridge University Press.

Quinn, P. K., Bates, T. S., Baum, E., Bond, T., Burkhart, J. F., Fiore, A. M., Flanner, M., Carrett, T. J., Koch, D., McConnell, J., Shindell, D., & Stohl, A. (2008). The Impact of Short-Lived Pollutants on Arctic Climate. ed. AMAP. Oslo, Norway: Arctic Monitoring and Assessment Programme.

Serreze, M. C., & Barry, R. G. (2011). Processes and impacts of Arctic amplification: A research synthesis. *Global and Planetary Change*, 77, 85-96.

Uttal, T., *et al.* (2002). Surface Heat Budget of the Arctic Ocean, *Bull. Amer. Meteor. Soc.*, 83, 255-275.

3.4. Integrate and Continue to Deploy a National Arctic Observing System and Promote International Cooperation to Create a Circumpolar Arctic Observing System

Lead Authors:
C. Nikoosh Carlo, NSF
Simon Stephenson, NSF
Martin O. Jeffries, ONR
Robert Sanford, NSF
Agency Partners:
DHS, DOE, DOI, EPA, NASA, NOAA, NSF, ONR

The Arctic Observing Network (AON) is being developed to provide data streams for parameters key to understanding the changing Arctic environment. Initially, a U.S. activity conceived by the Study of Environmental Arctic Change (SEARCH) group coordinated Arctic observing efforts expanded internationally during the International Polar Year 2007–2008, and these international efforts are now endorsed by the Arctic Council. Several documents and reports provide a vision of what an AON should be, as well as recommendations for network design, data approaches, input from local and other stakeholders, response to agency specific needs, and international partnerships (see end note). International collaboration has been developed further under the umbrella of the Sustaining Arctic Observing Network (SAON), including participation by the eight Arctic countries, indigenous organizations recognized as permanent participants in the Arctic Council, Arctic Council working groups, and non-Arctic countries and entities (Calder 2011). The U.S. contribution to the World Meteorological Organization (WMO) Global Atmosphere Watch provides another international cooperation mechanism.

For a list of Federal observing efforts see, *Arctic Observing Network: Towards a U.S. contribution to Pan-Arctic Observing* (IARPC 2007).

This section focuses on areas where interagency collaboration will greatly advance Arctic observing priorities and support the development of coordinated and integrated regional, national, and global observing systems. It will not attempt to review ongoing efforts at individual agencies. Suggestions for the general direction of network development are:

1. Agencies should establish key activities or priorities for Arctic research to support their mission. For example, NASA's comprehensive approach uses satellite, aircraft, *in-situ*, and model output to understand the causes of large, rapid changes in Arctic

sea ice, glaciers, ice sheets, land surfaces, permafrost, and atmospheric composition and chemistry. Those physical parameters strongly influence health and welfare of citizens throughout the pan-Arctic region, short-term weather patterns over the United States, and global change. NOAA's *Arctic Vision and Strategy* targets accurate, quantitative, daily to decadal predictions to support safe operations and ecosystem stewardship. Both priority areas are also included in the National Ocean Policy's Implementation Plan (http://www.whitehouse.gov/administration/eop /oceans/objectives) and will benefit greatly from interagency collaboration.

2. Residents of the North—individuals, communities, and their representative governments— need sound information and decision-making tools to inform adaptations to future conditions that will preserve, to the extent possible, their livelihoods, access to resources, and cultures.
3. Research should continue to build our understanding of the changing Arctic environment system and contribute to the design of an optimal observing network. From this, new priorities will be added; older ones may decline in importance or change focus.

The IARPC plans to focus on nine interagency observing network efforts over the next five years. A formal review and update of those priorities will occur every two years.

3.4.1. Facilitate Observing-System Design for the Arctic

Why Do This?

Long-term observations *in-situ*, from space, and by local people and communities have been vital to documenting changes occurring throughout the Arctic environment. Without remote-sensing observations from space, for example, the recent dramatic changes in Arctic Ocean sea-ice extent with melting and mass loss in the Greenland ice sheet with and changing tundra "greenness" would not have been detected and quantified quickly. The recognized need for a diverse set of pan-Arctic observations that would improve the value of predictive models spawned the AON.

The SEARCH articulated the need for AON in the first part of the last decade and continues today to refine thinking about the system via the SEARCH AON Design & Implementation Task Force (see the 2010 AON Program Status Report). More recently, the need was discussed in the National Academy of Sciences report, *Toward an Integrated Arctic Observing Network*. Components of the AON have been well-designed; the challenge now is how to link these effects across locations, regions, disciplines, or sectors. Progress in understanding the changing Arctic environmental system will be achieved by a variety of methods, including the use of coupled, numerical models that represent a regional Arctic system. Ideally, observations vital to model initialization, calibration, assimilation, and skill-testing are best provided by an optimal network. Optimizing an observing network is a design exercise that itself would employ numerical models to establish observing priorities and identify gaps, optimal observing sites, and observational needs, such as variables to be observed and the frequency and duration of observations. Thus, observing-system design and Arctic-system modeling are symbiotic and require the observing and modeling communities to work together to identify and develop synergies that will improve both approaches. Arctic regional models are discussed in more detail in section 3.5.

3.4.2. Assess Local-Resident Priorities for Addressing Change

Why Do This?

Polar amplification of global environmental change is forcing local Arctic residents to address change immediately and to identify new adaptation strategies while balancing ongoing social, cultural, health, and economic demands and expectations (see sections 3.6 and 3.7). Priorities for the Alaska Native communities include: human health well-being, subsistence rights and access, development of alternative energy, and rural dependence on fossil fuels, including predicting how climate change will alter their environment (see Alaska Federation of Natives 2010 and 2011 Federal Priorities and the Inuit Circumpolar Council Alaska Strategic Plan 2010-2014). Some observing networks should enable informed decision making by local Arctic residents.

3.4.3. Combine in-Situ and Remotely Sensed Observations of Sea Ice with Local Community and Traditional Knowledge

Why Do This?

The changing seasonality of sea ice will have profound climate, environmental, socio-economic, and political consequences. A sea-ice observing system that combines *in-situ* and remotely sensed observations with local community and traditional knowledge can meet the growing need for information about the state of the sea-ice cover at daily, weekly, monthly, seasonal, annual, and decadal temporal scales. As the Arctic Ocean moves toward a more maritime environment, the challenge will be to develop an optimal and integrated sea-ice observing system and derived information products and tools that will serve individuals and communities, industry, governments, and scientific research. The observing system will improve understanding of sea-ice processes and changing seasonality, operational forecasts, and longer-term predictions. Sea-ice processes, interactions, feedbacks, and modeling are discussed in more detail in section 3.1.

3.4.4. Conduct Long-Term Monitoring of Key Outlet Glaciers and Tidewater Glaciers

Why Do This?

Long-term monitoring is the only way to observe and study the evolving relationships between climate changes, perturbations at the ice/ocean interface, glacier flow, and mass loss.

Periodic measurements at key sites near Greenland and Arctic tidewater glaciers should capture glacier flow, local meteorology, ice mélange conditions, and oceanic conditions near the glacier front, in the fjord, and on the continental shelf.

Data collected should also provide a measure of the heat and freshwater transport into and out of key fjords to enable heat budget analyses and provide boundary conditions for ocean general circulation models (GCMs).

Essential variables include ice elevation, mass balance and flow speed, local meteorology, ocean temperature and salinity, and sea-ice conditions. These data will provide inputs for models of land ice loss described in section 3.5 as well as invaluable context for the study and validation of the linkages between key processes operating at vastly differing scales.

3.4.5. Monitor the Biological and Physical State of the Arctic Marine Environment

Why Do This?

Unprecedented ocean warming and seasonal thinning and retreat of sea ice are altering the biology of the Arctic Ocean and could potentially lead to ecosystem reorganization.

The international Arctic Council has initiated a program to monitor the biological state of the Arctic marine environment. In the United States, a DBO in the Pacific Arctic sector with emphasis on the Beaufort and Chukchi Seas is envisioned to provide biological and supportive environmental data at explicit regional sites.

Portions of the DBO represent the U.S. contribution to the Arctic Council's monitoring program. The DBO would use a collaborative international network of logistical support to track ongoing shifts in ecosystem structure concomitant with climate change. These are discussed more fully in section 3.1.

3.4.6. Assess the Effects of Clouds and Atmospheric Constituents on Surface Radiation Balance

Why Do This?

While the effects of Arctic warming are easily observed, the energy sources driving those changes are poorly understood. The surface-energy balance—the amount of energy from the sun versus the amount reflected or remitted back into space—is largely unknown due to the unusual effects of clouds and aerosols in polar regions (see section 3.3). The half-year-long polar night, the highly reflective surface of ice, and the unique atmospheric stability allow clouds and aerosols to enhance either warming or cooling depending on conditions.

Short-lived climate forcers, which include black-carbon aerosols, ozone, and methane, are also thought to enhance Arctic radiative forcing.

These and other complex cloud-aerosol processes contribute to some of the largest uncertainties in Arctic regional climate models.

3.4.7. Assess the Impact of Terrestrial Warming and Permafrost Thawing on the Carbon Cycle

Why Do This?

Recent estimates show that soils of high-latitude ecosystems store about 2,000 petagrams of carbon; a catastrophic release of carbon from the soils could be many times greater than anthropogenic carbon dioxide emissions.

Climate change scenarios predict the greatest magnitude of global warming will occur at high latitudes, and considerable observational evidence indicates recent warmer ground temperatures and permafrost thawing. The size of soil carbon pools and their sensitivity to temperature changes suggests a net loss of old-soil carbon to the atmosphere, causing a positive feedback to climate change.

Terrestrial ecosystems are discussed in more detail in section 3.2, and marine ecosystems, including greenhouse-gas fluxes, are discussed in more detail in section 3.1.

3.4.8. Improve Data Access

Why Do This?

Many data sets can be archived in the NASA's Earth Observing System Data and Information System (EOSDIS); NOAA National Data Centers (Geophysical, Oceanographic, and Climate Data Centers); NOAA's Arctic Ocean Observing System (AOOS); DOI/USGS Earth Resources Observation Systems (EROS) Data Center; DOE Atmospheric Radiation Measurement (ARM) and Next-Generation Ecosystem Experiments (NGEE) Data Archives; the NSF-supported Advanced Cooperative Arctic Data and Information Service (A-CADIS); the Geographic Information Network for Alaska at the University of Alaska; and the Exchange for Local Observations and Knowledge (ELOKA). Note that this is not an exhaustive list. The aim of these resources is to enable scientific discovery by allowing scientists to easily access, share, integrate, and work with data spanning multiple disciplines.

There remain several challenges that could benefit from interagency resources and expertise, which are detailed further in the Research Infrastructure Chapter 4, Table 3. These include, but are not limited to data exfiltration (unauthorized data transfer), data processing and archiving, establishing data standards, and data agreements that allow for interoperability of archives and support data discovery, analysis, and integration, and derived product development.

3.4.9. Engage Indigenous Observers and Communities in Monitoring Environmental Parameters

Why Do This?

Federal agencies supporting various components of the emerging circumpolar Arctic observation system, both nationally and internationally, will continue to work toward integrating community-based observation networks with physical and biogeochemical monitoring systems. Introducing human-focused components to observing and monitoring Arctic change has helped move the AON/SAON network far ahead of previous efforts. It is now widely recognized that engaging indigenous observers and communities in direct monitoring of various environmental parameters, such as weather, sea ice, coastal erosion, water resources, permafrost, marine and riverine resources, and terrestrial wildlife provides considerable value to both communities and researchers. There are strong positive examples such as EALÁT, a Reindeer Herders Vulnerability Network Study; the Bering Sea Sub-Network (BSSN); and the Sea Ice Knowledge and Use (SIKU) study—and similar projects begun under the IPY 2007-2008 program. During the next five years, agencies will explore the viability of converting such local initiatives into longer-term activities.

Timeframe

Long-term (>5 years) to achieve a sustainable AON; short-term (1-3 years) to make substantial progress on several high-priority areas.

Expected Outcomes

- Improved understanding of how the Arctic system is changing and its connections to global systems.

- Development of coordinated and integrated regional, national, and global observing systems.
- Informed decision making by local Arctic residents and other stakeholders.
- Easy access to, sharing of, integration across data spanning multiple disciplines.
- Coordinated monitoring network, including space and airborne missions, to measure and monitor the overall controls on Arctic land ice mass change, and the freshwater input into the North Atlantic.

Milestones

- Support diverse, multi-disciplinary observing teams that include representatives from state, local, and tribal governments, academia, the private sector, the international Arctic community, and other stakeholders (DHS, DOE, DOI, EPA, NOAA, NSF, ONR; FY2014).
  - U.S. Arctic Observing Coordination Meeting (DOI, NOAA, NSF, ONR; FY2012).
  - Arctic Observing Summit (NOAA, NSF, ONR; FY2013).
- Assess the state of the nine observing system/network themes and identify knowledge gaps and establish sites or regions for key observations (DHS, DOE, DOI, EPA, NOAA, NSF, ONR; FY2013).
- Develop an interagency (and international) planning document for an Arctic land ice monitoring system with focus on outlet and tidewater glaciers and their surrounding. (DOE, NASA, NOAA, NSF; FY2015).
- Develop action plans to implement an integrated design, including connections with other national and international observing systems, sustain current and planned operations, and use system models to identify observing contributions and needs for forecasting and design (all agencies; FY2016).

Science and Technology Gaps and Needs

New activities under this plan should be informed by ongoing efforts at the local, tribal, state, Federal, and international levels. Efforts should be made to review existing activities and create synergisms when goals and priorities align. Current research efforts would benefit from advisors from across Federal, state, local, and academic research organizations and industry to address AON topics. Such diverse perspectives on high-priority topics could quickly mobilize resources to advance areas of critical need.

Current inability to find some key datasets argues for a common portal to assist data discovery. Such a portal should be accessible and intuitive for local community members, educators, and the scientific community alike. Better access is needed to radar imagery over sea ice, glaciers, ice caps, ice sheets, and permafrost regions.

The ESA Sentinel series and the Canadian Space Agency Radar Satellite (CSA RADARSAT) Constellation will partly fill this gap but will depend on partnerships with overseas operations including free and open data sharing. Monitoring land ice loss will require continued development and deployment of an interdisciplinary observing system that is low-energy consuming, weatherproof, highly telemetrically efficient, and robust. In general there is great need for real-time data analysis systems capable of dealing with different data sources, types, and formats; and space and airborne campaigns that provide a continuous observational record.

References

Alaska Federation of Natives. (2010) Federal Priorities. http://www.nativefederation.org/AFNFedPriorities.php.

Arctic Observing Network [AON]. (2010). *Arctic Observing Network (AON) Program Status Report.* Results from the Third AON Principal Investigators (PI) Meeting, 30 November–2 December, 2009, Boulder, CO. http://www.arcus.org/search/aon.

Calder, J. (2011). Sustaining Arctic Observing Networks (SAON). In Krupnik, I., *et al.*, (Eds.), *Understanding Earth's Polar Challenges: International Polar Year 2007-2008.* Edmonton, Canadian: Circumpolar Institute.

Interagency Arctic Research Policy Committee [IARPC]. (2007). *Arctic Observing Network: Towards a U.S. Contribution to Pan-Arctic Observing.* http://www.nsf.gov/pubs /2008/nsf0842/index.jsp.

Inuit Circumpolar Council. (2011). *Inuit Circumpolar Council Alaska Strategic Plan 2010-2014.* http://www.iccalaska.org/servlet /content /other.html.

National Research Council. (2006). *Toward an Integrated Arctic Observing Network.* Washington, D.C.: The National Academies Press. http://www.nap.edu/catalog.php?record_id=11607.

Schlosser, P., Tucker, Warnick, W., & York, A. (Eds.). (2003). *Arctic Research Support and Logistics: Strategies*

and Recommendations for System-scale Studies in a Changing Environment. Fairbanks, Alaska: Arctic Research Consortium of the United States. pp. 81. http://www.arcus.org/logistics/03 report.html.

Solomon, S., Qin, D., Manning, M., Chen, Z., Marquis, M., Averyt, K. B., Tignor, M., and Miller, H. L. (Eds.). (2007). Contribution of Working Group I to the Fourth Assessment Report of the Intergovernmental Panel on Climate Change. Cambridge, United Kingdom and New York, NY, USA Cambridge University Press.

Study of Environmental Change [SEARCH]. (2005). *Study of Environmental Arctic Change: Plans for Implementation During the International Polar Year and Beyond.* Fairbanks, Alaska: Arctic Research Consortium of the United States (ARCUS). http://www.arcus.org/search/resources/ reportsandscienceplans.php.

End Note

Additional background on AON appears in:

- 2003 Arctic Research Support and Logistics: Strategies and Recommendations for System-scale Studies in a Changing Environment. ARCUS
- 2005 Study of Environmental Arctic Change (SEARCH): Plans for Implementation During the International Polar Year and Beyond. ARCUS
- 2006 Towards an Integrated Arctic Observing Network. NAS
- 2007 Arctic Observing Network: Towards a U.S. contribution to Pan-Arctic Observing. IARPC
- 2009 Arctic Observing Network (AON) Program Status Report. Results from the Third AON Principal Investigators (PI) Meeting, 30 November–2 December, 2009, Boulder, CO

3.5. Integrate Arctic Regional Models

Lead Author:
Mike Kuperberg, DOE
Agency Partners:
DOE, DOI, NOAA, NSF, ONR

Models of Earth's climate are mathematical tools for understanding climate processes and their feedbacks, as well as for predicting and projecting climate variability and change. A variety of models are being applied in the Arctic region for projecting future climate change, forecasting Arctic weather and sea-ice conditions, and understanding Arctic processes. Yet, due to large feedbacks, sparse observations, critical gaps in process understanding, differences in process representation, and high variability of vegetation, ice, and snow cover, significant uncertainty exists in simulations of Arctic changes. An integrated and focused effort to improve Arctic models would benefit understanding of ongoing processes, ability to project future Arctic changes, and informed use of those projections. Strongly coupling modeling and process-science research ensures that models incorporate state-of-the-science knowledge about critical systems.

Process studies (e.g., observations, experiments) can advance understanding in weak or unrepresented areas, and the resulting improved models can be used to guide field research/laboratory research and to inform future decisions.

Parameterization is a challenge of scaling—describing complex, smaller-scale processes as a simplified, mathematical representation in larger scale models. Close and improved collaboration between modeling and process scientists is needed to develop, test, and implement parameterizations.

The climate community has produced a number of documents that propose research directions; the following sections have made significant use of the *Science Plan for Regional Arctic System Modeling* (IARC), *Modeling and Predicting Arctic Weather and Climate* (Second *International Conference on Arctic Research Planning* (ICARP II), Science Plan 9, and the Arctic Climate Impact Assessment (ACIA).

3.5.1. Inventory Arctic Modeling Activities

Why Do This?

Significant efforts have been made to coordinate and share research on Arctic regional modeling. Rapid changes in the pan-Arctic region, however, call for even greater coordination of agency efforts and planned scientific research in Arctic change and modeling. Existing sources and portals that provide a base for comprehensive information and data coordination and sharing include:

- National Snow and Ice Data Center (NSIDC), Boulder, CO, which includes NASA's Distributed Active Archive Center (DAAC) for all cryosphere related data and NSF's Advanced Cooperative Arctic Data and Information Service (ACADIS).
- The Arctic Portal established as part of the International Polar Year activities and currently maintained by Iceland and various Arctic Council Working Groups

- ARM Climate Research Facility Archive (http://www.archive.arm.gov/armlogin/login.jsp)
- Earth System Grid (http://www.earthsystemgrid.org)
- Polar Hydrography Center, University of Washington
- U.S. National Academies Polar Research Board
- Alaska Ocean Observing System (http://www.aoos.org/)
- North Slope Science Initiative (http://northslope.org/)
- Arctic Ocean Model Intercomparison Project workshops
- NASA Catalog of Earth Science System Components (http://www.asd.ssc.nasa.gov/m2m)
- Community Earth System Model (CESM) Polar Climate Working Group (http://www.cesm.ucar.edu/working_groups/Polar)
- NSF's Arctic Long Term Ecological Research (LTER) sites (http://www.lternet.edu/)
- Arctic Landscape Conservation Cooperative (http://www.arcticlcc. org)
- Scenarios Network for Alaska Planning (http://www.snap.uaf.edu/)

Access to and broader use of existing models, data systems, and information portals would better coordinate and integrate existing agency efforts and improve dissemination of data and information about Arctic research activities. An inventory of modeling activities, process research, and data sources would provide a basis for defining collaborative programs and coordinated campaigns for advancing Arctic modeling. Cataloging the existence and nature of activities, such as model inter-comparisons and benchmarking, should be included in such an inventory.

Timeframe

Near-term (1-2 years)

Expected Outcomes

- New efforts to intrinsically couple modeling and process research activities.
- More rapid integration of knowledge into models.
- Improved capability for model evaluation to inform process research.

Milestones

- Conduct and disseminate a survey of Federal Arctic modeling efforts (DOE, NSF; FY2013).
- Evaluate the results of the survey and identify opportunities for collaborative development and/ or joint campaigns (DOE, DOI, NASA, NOAA, NSF, ONR; FY2014).

Science and Technology Gaps and Needs

A comprehensive bibliographic database of results from international entities [such as the International Arctic Science Committee and the World Climate Research Program's (WCRP) Climate and Cryosphere Program] would provide needed context for how modeling relates to ongoing studies.

3.5.2. Encourage Coordinated Approaches That Better Represent Arctic Processes in Earth-System Models

Why Do This?

Global Earth system models (ESMs) have become increasingly robust, accurate, and comprehensive. Large differences exist in the Arctic systems simulated by different ESMs and many processes are either absent (dynamic ice sheets, ice ecosystems), poorly represented (multi-phase clouds, permafrost, vegetation) or unresolved (ice fracturing, ice shelves, ocean eddies) in Arctic models. Feedbacks in the Arctic are also particularly strong and sensitive; thinning sea ice is very sensitive to both atmosphere and ocean forcing and becomes highly variable with subsequent changes in albedo and other feedbacks; biogeochemical processes that control the rate and nature of carbon release are suspected to lead to dramatic effects on atmospheric carbon concentrations from thawing permafrost; and ecosystem changes may have direct (albedo) and indirect (ecosystem services) impacts.

Such processes and feedbacks control both regional Arctic and, indirectly, global climate variability. Their realistic representation requires models with very high spatio-temporal resolution as well as detailed and often long-term observations to improve parameterizations and model verification. High-resolution models of Arctic processes will subsequently need to be parameterized and/or scaled for use in current ESMs. International coordination of efforts and cost sharing are important while maintaining model diversity for more robust evaluation of a variety of approaches.

Timeframe

Mid-term (3-5 years) for many model improvements already in the pipeline; long-term (5-10 years) for next-generation models. The Earth System Modeling Framework (ESMF) (http://www.earthsystem modeling.org/) and the Earth System Grid Federation (ESGF) (http://esgf.org/) are currently making models and their output accessible to the larger community but further coordination is needed.

Expected Outcomes

- Improved interagency coordination through dedicated multi-agency campaigns on specific Arctic processes.
- Continued development of common modeling frameworks, data and meta-data standards to facilitate increased sharing of model components, simulation, and observational data.
- Continued improvement of process representation of the Arctic region in Global Earth system models.

Milestones

- Identify critical Arctic processes for dedicated field and modeling campaigns across agencies (DOE, DOI, NASA, NOAA, NSF, ONR; FY2015).
- Coordinate Federal activities to develop, implement, and test improved parameterizations of Arctic physical processes and feedbacks (DOE, DOI, NOAA, NSF, ONR; FY2015).
- Develop standardized model components, meta-data, and data products (DOE, DOI, NOAA, NSF; FY2016).

- Conduct model inter-comparisons to foster collaboration among modeling groups and identify high-priority Arctic model improvements (DOE, NOAA, NSF; FY2018).
- Review report from the third session of the WMO Executive Council Panel of Experts on Polar Observations, Research, and Services on the development of the Global Integrated Polar Prediction System (NOAA, NSF; FY2012).

Science and Technology Gaps and Needs

While some coordinated efforts exist, such as the Arctic Ocean Model Intercomparison Project (AOMIP) and Climate Process Teams, additional opportunities exist to leverage and coordinate current and planned Arctic research investments. Enhanced mechanisms for interagency collaboration are therefore needed.

3.5.3. Build Arctic and Subsystem Models for Coupling with Regional and Global Approaches

Why Do This?

Regional models of the Arctic are important tools that enable explicit representation of fine-scale atmospheric, oceanic, terrestrial, and cryospheric processes at scales where they occur.

Regional Arctic models can also take advantage of the growing number of data streams and process understanding that is emerging from mechanistic studies and field and laboratory observations. They can also highlight the most critical and sensitive processes and parameters for inclusion in regional and global models.

Regional models are required to fully incorporate the complex interactions between sea ice, ice sheets, cold oceans, regional climate, permafrost stability, and their integrated influence on the regional and global carbon cycle and energy balance.

Ultimately, the understanding obtained from these models can be incorporated into global models either through coupling/nesting or by creating appropriate parameterizations.

Global Earth system models will be the primary tool for exploring climate change and the Arctic feedbacks in the global climate system.

Timeframe

Mid- to long-term (3-10 years). Focused development of new subsystem models and interaction with experimentalists (>5 years); coupling and evaluation within regional and later in global models (5-10 years).

Expected Outcomes

- A hierarchy of models from process to regional scales.
- Inclusion of Arctic high-resolution physical, biological, chemical, and social subsystem model components within a regional Arctic climate system model.
- Improved representation of the Arctic system in Earth system models through nesting of regional models or development of parameterizations.

Milestones

- Develop and evaluate stand-alone subsystem components of the Arctic system, incorporating mechanistic processes derived from experiments and/or observations (DOE, DOI, NSF; FY2014).
- Couple, test, and validate the aforementioned models against observations of subsystem components within a regional Arctic climate-system model (DOE, DOI, NOAA, NSF, ONR; FY2016).
- Couple and evaluate Arctic subsystem components within Global Earth system models (DOE, DOI, NOAA, NSF, ONR; FY2017).

Science and Technology Gaps and Needs

Observational data are needed to constrain and rigorously evaluate individual subsystem model components and multi-component interactions and dependencies. Advanced *in-situ* and remote platforms are needed for coordinated process studies, especially for difficult observations such as under-ice sampling and communication, melting out/freezing in, long-time/distance endurance, high-resolution, large-scale, and long-term coverage. Also, community participation in model development and data collection could be better supported by organizational structure as would common, coordinated Arctic data/model distribution center(s) for community posting and use of models.

3.5.4. Develop Models of Arctic Land Ice Mass Loss, Connections to Ocean and Atmospheric Variability, and Implications for Sea Level

Why Do This?

Predicting future changes in ice sheet mass and the impacts on sea level, ocean circulation, and climate will require improved understanding of ice-sheet and ice-ocean processes. Data from process-oriented studies, targeted campaigns, and long-term monitoring systems must be integrated into large-scale circulation and Earth-system models. The spread of current projections of sea level rise contributions from Greenland by 2100, from 0.06 to 0.54 m, with an added 0.34 m from glaciers and ice caps, gives a crude measure of current uncertainties.

Timeframe

Mid-term (3-10 years)

Expected Outcomes

- Implement and test physically based parameterizations of unresolved processes: key physical processes operating on scales of 100 m or less need to be incorporated into Earth system models using a suite of parameterization techniques.
- Comprehensive, well-structured, and sophisticated databases and data formats which allow rapid access and optimal use of the hard-won data.
- Coupling of the various components of the Earth system models: representation of feedbacks between ice sheet variability and the large-scale ocean/atmosphere circulation or other climate system components through interactive (two-way) coupling between ice sheet and climate models or components.

- Coupling and accounting for Earth gravity/geodetic responses in models: glacial isostatic adjustment (GIA) from visco-elastic mantle response to unloading and gravitationally self-consistent sea-level solutions are crucial in assessing regional sea level change as a consequence of ice sheet mass loss; gravity models of improved accuracy are needed and results fed back into Earth system models (which currently neglect these effects).
- Model testing, analysis, and intercomparison: a hierarchy of modeling approaches serves as a quantitative basis for model assessment and identification of systematic biases; a feedback loop links large-scale model-data biases, e.g., subpolar gyre hydrographic properties, to those at key locations (fjord exits and glacier termini), and to discrepancies of parameterized versus observed processes.
- Use of models to optimize observing-system development: these studies will assess which processes have the strongest impact on constraining ice mass loss, and where, with what accuracy, and at which frequency these should be sampled.

Milestones

- Support surveys of existing modeling capabilities by participating agencies that tackle the ocean/land ice coupling problem, including parameterization and moving geometry approaches (DOE, NOAA, NSF; FY2014).
- Help coordinate the implementation of multi-disciplinary test suites and scenarios for idealized and realistic model verification, validation, and intercomparison (a combination of ISOMIP and SeaRISE but extended to the coupled problem); foster exchange of codes, inputs, experiences (DOE, NSF; FY2014).
- Translate results from process studies into parameterizations for use in Earth System Models (DOE, NOAA, NSF; FY2015).
- Development of new-generation modeling approaches into the coupled forward and inverse problem (DOE, NOAA, NSF; FY2018).
- Completion of observing-system studies using improved models to further inform observing-system design (DOI, NOAA, NSF; FY2023).

Science and Technology Gaps and Needs

- Numerical methods are needed that can deal with processes acting on a wide range of spatiotemporal scales, from sub-centimeter boundary layers to ocean basin-scale (1000 km) coupled atmosphere/ocean dynamics. Three types of complementary approaches may be pursued: (1) process-resolving simulations to develop process parameterizations; (2) feedback and testing of these parameterizations in large-scale Earth system models; and (3) in parallel with (1) and (2), the development of new computational tools (unstructured and/or adaptive meshing, multi-grid, multi-physics approaches) to enable seamless simulations. Algorithms are required that can efficiently exploit emerging massively parallel computer systems and integrate different system components. Yet, simulations also need to produce diagnostic output that is useful for scientific analysis.
- Dynamical feedbacks through moving geometry (glacier retreat) represent a new computational challenge in climate modeling; coupled ocean-glacier simulations that aim to capture dominant glacier stress perturbations through glacier un-grounding and retreat need to represent the changing interface geometry.

- Observational sparsity requires the continued development and use of formal inverse/estimation methods in a coupled model/data framework; merging of diverse observations and models is needed to infer optimal state and parameters and to assess remaining uncertainties.

3.5.5. Increase Arctic Model Resolution to Improve Prediction and Inform Future Research and Observations

Why Do This?

Accurate simulations of the Arctic require the ability to resolve features and processes at small spatial and temporal scales. For example, Arctic Ocean eddies, cloud processes, ice fracturing, ice-shelf dynamics, ecosystem/vegetation changes, and hydrology all occur on kilometer scales or smaller. Through data synthesis and integration with other models, high-resolution regional models can guide future observations and process studies while providing a large-scale and long-term picture of Arctic system variability. Variable resolution approaches are now being implemented and show great promise for enabling high-resolution Arctic models in a global model framework. Similarly, alternative time integration techniques are being explored to capture multiple time scales consistently and accurately.

Process understanding, data sets, and model representations are needed at small spatial and temporal scales. High-resolution model output is also commonly required by stakeholders, especially for impact assessment and mitigation/adaptation studies. While high-resolution global models are beginning to reach these spatial scales, continued use of regional, stand-alone, and nested models will be needed to provide ensemble simulations at the required resolutions. Similarly, fast processes related to surface exchange and boundary layer processes must be represented, as well as decadal modes of variability and longer processes like ice-sheet melting, ecosystem evolution, and ocean thermohaline circulation—all of which have significant impacts on the Arctic. Results from those simulations can inform process research direction and observing system design (see section 3.4). Already, high-resolution simulations have been used to help design large-scale ecosystem studies in Arctic permafrost and float experiments in the Southern Ocean.

Timeframe

Mid-term (3-5 years) for improved Arctic climate-system models and variable-resolution models for focused Arctic simulations

Expected Outcomes

- Improved and detailed forecasts of Arctic climate change to address stakeholder needs.
- Arctic simulations at fine-enough resolution to inform experimental/observational design and field campaigns.

Milestones

- Conduct ensemble simulations of future Arctic climate change at kilometer spatial scales (DOE, DOI, NSF; FY2018).
- Engage the Arctic (and Earth system) modeling community in planning and designing future field campaigns (DOE, DOI, NOAA, NSF, ONR; FY2012).

- Provide mechanisms for rapid access to processed (quality controlled, formatted/gridded) observational data sets for model-data inter-comparison (DOE, NOAA, NSF; FY2015).

Science and Technology Gaps and Needs

Much observational data are not at high enough resolution to validate kilometer-scale models or to incorporate into model-data inter-comparisons. Important gaps in process understanding limit full representation of the Arctic in coupled models. Advanced numerical techniques are needed for multi-scale time integrations. High-performance computing and technical support are needed for model integration thereby enabling co-design of model algorithms and/or new modeling approaches to carry out high-resolution simulations on new computing architectures.

3.5.6. Use Insights from Models to Inform Process Research; Use Process Research to Evaluate and Improve Models

Why Do This?

Models can be used to simulate processes that are difficult to observe and, thus, poorly understood.

Analyzing models can highlight areas where new observations or studies will improve representation of processes in models and ultimately benefit predictions about Arctic climate change.

Modelers are eager for robust data sets to evaluate and test their products, and experimentalists are collecting observations and data that can be incorporated into models. The two communities, however, do not often interact directly.

The disconnect among modelers, experimentalists, and observers has led to gaps in both the completeness of data for good prediction and in critical process representation in models.

Coordinated research efforts are particularly needed in the Arctic where physical conditions and remote locations make it difficult to conduct process research and where different models currently provide very different representations of Arctic parameters and/or processes.

Great opportunity exists to rapidly improve the design of Arctic experiments and observations, which would enable enhanced parameterization of subsystem, regional, and global models. Such experiments and observations also could identify and address structural uncertainty in models by including coupled experimental and predictive science components to identify gaps and needs simultaneously for both data and models.

Timeframe

Near- to long-term (1-10 years). Current field-research efforts and laboratory process studies are already being informed to some extent by model analyses, but more comprehensive and integrated model and field campaigns will be needed.

Expected Outcomes

- Improved understanding and model representation of processes specific to the Arctic system.

- A new cycle of systems model development and improvement, incorporating integrated field campaigns to develop refined parameterizations for Arctic process representations in models.
- More efficient use of research funds through enhanced collaboration among process researchers, observationalists, theoreticians, and modelers.
- Rapid increase in development of robust data sets, process understanding, and improved models and predictive skill.

Milestones

- Publish SeaRise ice sheet model intercomparison results (DOE, NASA, NSF; FY2013).
- Analyze model output to determine future needs for data collection and process studies (DOE, DOI; FY2014).
- Design and implement integrated modeling and field campaigns focused on specific high-priority processes to improve process understanding and representation in models (DOE, DOI, NASA, NOAA, NSF, ONR; FY2012).
- Develop, test, and evaluate new/improved parameterizations (DOE, NOAA, NSF, ONR; FY2015).
- Plan and conduct interagency and intra agency conferences, workshops, and campaigns designed to bring process researchers and predictive scientists together to solve Arctic grand challenges (DOE, DOI, NASA, NOAA, NSF, ONR; FY2015).
- Develop funding solicitations that require integrated process-prediction research approach (DOE, DOI, NSF, ONR; FY2013).

Science and Technology Gaps and Needs

Closely connecting process and modeling science is needed to rapidly improve and validate our ability to project future Arctic conditions. Many processes in Arctic systems are not well understood and could be targets for focused research. Examples of key areas of research needs include:

1. Ice thickness distribution and ice production due to ice deformations;
2. Multi-phase ice, brine channels, melt ponds, and water/brine transport;
3. Ice-sheet and ocean interaction, ocean circulation in tidewater fjords, and property exchange over the sill;
4. Time-dependent evolution of surface, englacial, and subglacial hydrology for the Greenland ice sheet;
5. Ocean circulation and water-mass formation in ice-free Arctic conditions;
6. Ocean estuary/fjord subsystem model component for use in climate models;
7. Land/submarine permafrost distribution and interaction with atmosphere and/or ocean;
8. Permafrost hydrology and changes in morphology and sheet flow and riverine transport of surface water, sediments, nutrients, and contaminants;
9. Arctic biogeochemical cycles within permafrost, sea ice, atmosphere, and ocean, including carbon, methane, ice algae, aerosol-cloud interactions, aerosol deposition on ice;

10. Arctic mixed-phase clouds and other cloud microphysical processes, including ice nucleation;
11. Surface exchange processes at ice/ocean, ice/atmosphere, and ocean/atmosphere boundaries;
12. Rates and ranges of change for plants, animals, and ecosystem function; and
13. Permafrost-soil-vegetation interactions.

3.5.7. Integrate Arctic Climate-Model Results with Observational Validation and Verification to Understand the Principal Drivers and Uncertainties of Arctic Climate Changes

Why Do This?

Unified, regional Arctic climate-system models are ideal tools to integrate data and information across multiple components of the Arctic system. Such models may also be applied to advanced, probabilistic decadal projections. Models must be verified against analytic test cases and reference solutions and validated against the historical record. Ensemble simulations are required to sample the solution space for uncertainty quantification as well as to separate natural or internal modes of variability from externally forced changes.

As with all aspects of climate modeling, critical evaluation needs to continue for Arctic regional models. Our ability to validate model projections (and to describe the uncertainties associated with those projections) is critical to our ability to use modeling results. Model validation contributes to the optimal synthesis and integration of limited process understanding and observational data to advance Arctic science, including interdependencies across the Arctic system components.

Timeframe

Mid- to long-term (3-10 years)

Expected Outcomes

- Projection of future Arctic climate change and its impacts including variability estimates.
- Determination of key principle drivers of Arctic climate variability and trends.
- Integration of comprehensive analyses of causes and effects of Arctic climate change.

Milestones

- Coordinate model experiments and inter-comparisons to critically evaluate regional model results against observations (DOE, NOAA, NSF; FY2014).
- Develop and implement standards for gridded observational data sets (DOE, NOAA, NSF; FY2014).
- Implement a common data portal for both observational data and common model experiments and projections (DOE, DOI, NSF; FY2018).

Science and Technology Gaps and Needs

Improved diagnostic measures suitable for both observations and model data are needed, as well as statistical techniques for uncertainty quantification. Improved coverage and long-term monitoring of the Arctic will generate better knowledge of present climate and initial conditions for future projections. New techniques for filling in sparse observational data from the historical record will improve model validation.

Model improvements and additions should occur as a continuous, iterative cycle. Enhanced computational and data infrastructure would enable improved integration of currently disparate data streams and the ability to exercise increasingly complex (and, therefore, realistic) model representations of Arctic systems.

The enhancements include improved data archives and distribution, advanced informatics, and access to high-performance computing for high-resolution ensemble simulations.

3.6. Assess Strengths and Vulnerabilities of Arctic Communities Facing the Impacts of Climate Change and Assist in Developing Adaptation Strategies and Tools to Maximize Sustainability, Well-Being, and Cultural and Linguistic Heritage

Lead Authors:
William Fitzhugh, SI
C. Nikoosh Carlo, NSF
Igor Krupnik, SI
James Partain, NOAA
Agency Partners:
DOI, DOS, EPA, NSF, NOAA, SI

Rapid Arctic change is forcing residents to adapt to new conditions created by environmental change and diverse socio-economic stressors. Age-old traditional responses, such as diversification of natural-resource harvesting, relocation, and dispersal or concentration in the resource-rich niches, offer valuable lessons, but by themselves may be inadequate to fully address challenges such rapid changes and modern economic transitions present. New community-based participatory research to identify regional and local vulnerabilities and adaptation tools contribute to the growing body of knowledge needed for decision-support science and policy recommendations. Understanding the impacts of climate change could be better served by adding base—or "bottom-up"—assessments of the current drivers of social well-being to computer-generated models and other top-down scenarios used in ecosystem modeling. Focused studies of major social parameters of change are under way that are national and international in scope. Integrative multi-disciplinary programs, like SEARCH, that seek to balance societal needs with research priorities advanced by rapid environmental change can help explore effective ways of engaging communities and Tribes in the issues of utmost public urgency, such as resource development or management of living resources. *Arctic Social Indicators II* and *Arctic Human Development Report II*, projects endorsed by Arctic Council, will emphasize the key role of local communities in both basic-science discovery and in creating adaption tools that are practical and efficient at the local, regional, and broader circumpolar levels. Knowledge about thresholds and prospective breaking points, as well as the inherent strengths in the social fiber that affect community resilience, would be invaluable in formulating Federal agency policies and in forging collaborations with diverse local stakeholders. The overall scope of new research should be broad and encompassing, but the informative case studies are to be conducted at the community level. Melding the historic effects of cash and natural-resource-harvesting economies with added stressors associated with modernization and ongoing climate change is challenging but necessary. The emphasis should be to provide Arctic residents with basic scientific knowledge necessary for community leaders to develop sustainable pathways for successful adaptation amid rapid environmental change and a variety of other stressors—while juggling diverse Federal, state, and local interests. Providing such knowledge will help preserve cultural vitality and improving health and overall well-being for the long-term. The need to support tribal communities so they have representation in climate change-related activities and can develop and implement climate change strategies has also been recommended by the Climate Change Adaptation Task Force in its 2011 Progress Report.

3.6.1. In Collaboration with Local Communities, Develop Methods for Assessing Community Sustainability and Resilience and Determine the Efficiency of Current Adaptation Strategies

Why Do This?

Native communities have thrived in the Arctic for millennia, but today's challenges are occurring at a pace faster than ever and may be beyond the capacity of traditional adaptation strategies. As primary stakeholders that will be affected by climate change, Arctic communities and Tribes need useful, access-ible, location-specific information on changes that are occurring and how those changes will impact human well-being. There is an urgent need for effective methods to assess community sustainability as northern residents face various socioeconomic challenges, changing local infrastructure needs, and high transportation and living costs. Traditional ecological knowledge should be actively incorporated in future planning, and more research is needed to identify key indicators of successful adaptation at local and regional scales.

Timeframe

Near-term (3-5 years)

Expected Outcomes

- Inform new understanding on how climate change, past and present, has been/is being met by Arctic communities and how communities have adapted.
- Improve research and assessment tools for studying the sources of community vulnerability to climate and environmental change and implications for community adaptability developed in partnership with local communities and organizations.

Milestones

In collaboration with other Arctic nations:

- Identify and develop a database on past and current adaptation strategies used by Arctic communities to combat climate change impacts (DOS[10]; FY2013).
- Determine which strategies have been most successful (DOS; FY2013).

- Document unintended consequences of previous strategies and responses to change (DOS; FY2013).
- All will be explored by the Adaptation Actions for a Changing Arctic Report (DOS; FY2013).

Science and Technology Gaps and Needs

Limited work has been done to identify what Arctic communities have done, or are currently doing, to adapt to the effects of climate change. Even less is known about the efficacy of those adaptation strategies. Research is needed to, first, identify the suite of climate-change adaptation strategies currently being (or that have been) used and, second, to analyze the efficacy of those strategies. Particular attention should focus on both community specific strategies and those that were applied more broadly at the regional, national, and circumpolar scales. All research needs to be conducted as joint efforts and in close collaboration with local Arctic communities.

3.6.2. Identify the Current Vulnerabilities of Arctic Communities and Ecosystems to Climate Change and Explore Their Interactions with Socio-Economic and Other Stressors

Why Do This?

The Arctic region is warming quickly. The impacts of this change on Arctic communities are occurring in conjunction with other stressors, such as high cost of living, infrastructure maintenance, lack of employment opportunities, and shortage of resources for community development (see Section 3.7 for discussion of health-related challenges). These stressors need to be identified and quantified in pragmatic terms. Successful adaptation measures will require attention to multiple stressors simultaneously as well as close engagement with communities to empower self-sufficiency, initiate smart development, and build adaptive capacity.

Timeframe

Near-term (1-3 years)

Expected Outcomes

- Assist local communities in better understanding current vulnerabilities to climate change.
- Strengthen or create new partnerships among local stakeholders and resource managers to assess the vulnerability of individual Arctic communities and develop locally based adaptation and more general mitigation strategies.
- In collaboration with local stakeholders, develop new understanding of how changes in Arctic ecosystems and climate will affect natural-resource harvesting practices.
- Improve public understanding of how the effects of climate change interact with socioeconomic stressors in the Arctic.
- Advance engagement of Federal agencies with local communities in sharing information on management and use of terrestrial and marine resources, and support empowerment for the co-creation of knowledge.

- Community profiles that highlight impending vulnerabilities and indicators of resilience.

Milestones

- Establish research and community collaborations aligned with local priorities and needs, including in planning, data collection, conceptualization, and interpretation of research results and recommendations (NSF; FY2013).
- Assess vulnerability of Arctic communities and ecosystems to climate change and socioeconomic stressors:
  - Environmental Studies Program Social Indicators in Coastal Alaska-Arctic Communities Study (BOEM; FY2014).
  - Arctic Sustainability (ArcSEES) initiative (BOEM, EPA, NSF, USFW, USGS, and a consortium of French science funding agencies; FY2013).
- In collaboration with other Arctic nations, develop a standardized set of quantifiable socioeconomic indicators of vulnerability in the Arctic Council's Arctic Social Indicators II Study (DOS; FY2012).

Science and Technology Gaps and Needs

Research on the vulnerabilities and indicators of resilience of individual Arctic communities to climate change is limited. Thus, there is a need to develop quantitative measures of vulnerability and resilience to climate change for coastal communities, which can be compared with concurrent socioeconomic stressors. More-objective analysis will improve the content and quality of socioeconomic-impact assessments associated with policy actions and marine resource-management decisions. Such research will enable affected communities to develop strategies for best responding to the challenges and opportunities they face.

3.6.3. Develop Projections of Future Climate Scenarios and Demographic Conditions to Forecast Potential Strengths and Weaknesses of Human and Ecological Systems in the Arctic

Why Do This?

Given current global environmental changes, the future will not be like the past; scenarios of future conditions help define alternate projections of environmental and socioeconomic states and—with identified uncertainties—will be valuable to adaptation planning.

Timeframe

Long-term (>5 years) Expected Outcomes:

- Improved models that allow communities and resource managers to better assess potential community and ecosystem-level vulnerabilities to medium- and long-term climate scenarios as well as the relative benefit of specific actions to address those vulnerabilities at the decadal scale.

Milestones

- In collaboration with other Arctic nations, develop a standardized set of socio-economic indicators to measure future community resilience (2020 and 2030 projections), including input of local resources, population fluctuations and migration, communication networks, and capacity to adapt via the *Arctic Council's Arctic Social Indicators II Study 2012* and *Arctic Resilience Report* (DOS; FY2015).

- Link climate models with projections of ecological and socio-economic change that include community dependence on harvesting local food sources (BOEM, DOE, NOAA; FY2017).
- Test existing scenarios of the status of certain wildlife species of value to local consumers, according to available climate change models (USFW; FY2017).

Science and Technology Gaps and Needs

Knowledge is lacking about how climate change might disproportionately impact communities and the ecosystems that they depend on for survival. A comprehensive research plan needs to be developed to collect the suite of information needed to model forecasted impacts. Once such models are created, adaptation strategies that respond to those impacts can be developed with local communities and various state and Federal agencies that can aid in moving explicit goals forward.

3.6.4. Assist Arctic Communities in Documenting, Revitalizing, and Strengthening Indigenous Languages and Cultural Heritage

Why Do This?

Arctic communities have long faced threats to their indigenous languages and cultural continuity. New technologies, communication, education, and digital media are primarily in non-indigenous languages. At the same time, the digital era offers new channels for distant education, online publication, and broad dissemination of cultural materials. It is also creating new challenges, as most traditional cultural, subsistence, and language skills are still being transmitted by direct contact within families and communities. The impact of these and other new forces on the Arctic's indigenous people is poorly understood. Federal institutions need to develop a unified strategy or policy for assisting Arctic residents with 21st century challenges to their cultural well-being. Many communities are eager to address the progressive loss of traditional knowledge and language and are willing to work in partnership with Federal and local agencies to develop new strategies to preserve and use their unique cultural heritage.

Timeframe

Mid-term (3-5 years) Expected Outcomes:

- Assess the existing Federal and State of Alaska resources that support Arctic people's cultural heritage and ways to expand collaboration with indigenous communities in language, knowledge, and heritage preservation.
- Enable informed decision making by users—local programs, cultural institutions, schools, branches of local governments and communities—for collaborative actions in support of indigenous cultural heritage, languages, and identities.
- With local communities develop new programs and templates for cultural preservation, including language revitalization.

Milestones

- In concert with local communities, strengthen partnerships between researchers, Alaska Native organizations, Federal, State, and non-governmental organization (NGO) entities through strategic projects, workshops, and conferences (DOE, DOI, NEH[11], NSF, SI).
    - Beringia Days Annual Event (DOI's National Park Service; FY2013).
    - Documenting Endangered Languages Program (NEH, NSF; FY2013).
    - Inuit Studies Conference (SI; FY2012).
- Develop tools that Arctic communities can use to more effectively support indigenous languages, traditional ecological knowledge, and natural resource harvesting activities (NEH, NSF, SI; FY2014).
    - Assessing the Vitality of Arctic Indigenous Languages - Research Development Workshops (NSF; FY2012).
    - Recovering Voices Program ongoing outreach (SI; FY2014).
- Create community profiles that highlight continuity of indigenous languages and knowledge systems (NEH, NSF; FY2014).
- Identify and strengthen Federal, State, and local efforts related to indigenous languages, traditional knowledge, and cultural heritage (DOI, NSF, SI; 2014).
- Develop research programs to monitor the status of indigenous languages and knowledge systems and to better understand factors affecting language and cultural resilience (NEH, NSF, SI; FY2016).

Science and Technology Current Gaps and Needs

Arctic communities have been and will continue to adapt to new technologies, communication, and preservation tools to sustain their cultures and languages. The strengths and threats from the new electronic and media era to indigenous languages have not been adequately studied, nor is a reliable assessment of the opportunities offered by new technologies available to guide Federal and state policies. Digital technologies are being increasingly used in support of indigenous languages and knowledge systems, but many critical issues, such as data management, data sharing, compatibility, and local resources remain unresolved. Research is needed to identify key components necessary for the application of new digital technologies and for collaboration of the many players at the Federal, state, local, and community levels.

3.7. Understand Factors That Impact Human Health in the Arctic, Including Infectious and Non-Communicable Diseases, Climate Change, Environmental Contamination, and Behavior and Mental-Health Disorders

Lead Authors:
Alan Parkinson, CDC
Marya Levintova, NIH
Agencies Partners:
CDC, EPA, IHS[12], NIH, USARC

Although health and survival of Arctic indigenous peoples have improved over the past 50 years, important disparities remain in life expectancy, infant mortality, and leading causes of death when compared with their respective national populations. Life expectancy of the indigenous peoples of Alaska, northern Canada, and Greenland is lower than that of the general populations of the United States, Canada, and Nordic countries. Similarly, infant mortality among indigenous segments of those populations is higher than that of the comparable national populations. Meanwhile, mortality rates for heart disease and cancer—once much lower among the indigenous populations of the United States, Canada, and northern European countries—now mirror their respective national rates. In addition, indigenous populations of Alaska, Canada, and Greenland have higher mortality rates for unintentional injury and suicide believed to result from a complex disorder of depression, alcoholism, child abuse, and other substance abuse (Levintova *et al.* 2010). Other health concerns of those groups include a high prevalence of infectious diseases such as hepatitis B and C, *Helicobacter pylori*, infant respiratory syncytial virus (RSV), tuberculosis, and sexually transmitted diseases, as well as heath impacts associated with exposures to environmental pollutants, rapid economic change and modernization, and climate change (Young and Bjerregaard 2008).

The majority of DHHS-supported research in the Arctic is carried out within Alaska in collaboration with scientists at the University of Alaska, the State of Alaska, and the Alaska Native Tribal Health Consortium (ANTHC). Most activities are funded by individual NIH institutes and/or CDC centers; however, some activities receive support from non-DHHS agencies, such as the EPA and U.S. Arctic Research Commission. The CDC maintains the Arctic Investigations Program (AIP), which is a field station located on the ANTHC campus in Anchorage. AIP provides a platform for collaborative biomedical and prevention research focused on improving the health of Alaska Native population and promotes circumpolar health through linkages with CDC centers, the International Union for Circumpolar Health, and the Arctic Council's Sustainable Development Working Group's Arctic Human Health Experts Group (Parkinson 2010). CDC also maintains the Alaska Pacific Regional Office of the National Institute for Occupational Safety and Health, which conducts occupational safety research to reduce hazards in the Arctic.

Supporting further multi-agency research will contribute to knowledge of factors affecting human health in the Arctic and lead to development and implementation of prevention and control and successful treatment strategies that will greatly improve the health and well-being of Arctic residents.

3.7.1. Continue to Expand Circumpolar Surveillance and Research for Infectious Diseases, Non-Communicable Diseases, Trauma, Injury, Sanitation Services, and Indoor Air Quality to Help Prevent Morbidity and Mortality

Why Do This?

Effective surveillance can facilitate timely control of disease outbreaks, inform public health decisions on research and resource allocation, and provide data to maximize prevention and control strategies. Population-based surveillance of diseases of concern is conducted by public-health agencies in the United States, Canada, Greenland, Iceland, Norway, Finland, Sweden, and the Russian Federation. The International Circumpolar Surveillance (ICS) of Emerging Infectious Diseases was established in 2000 as an Arctic Council Sustainable

Development Working Group project. Linkage of the individual national systems has created a network of hospitals, public-health agencies, and reference laboratories throughout the Arctic to collect, compare, and share uniform laboratory, epidemiological, and research data on infectious diseases and assist in forming prevention and control strategies (Parkinson *et al.* 2008).

Cancer research depends upon population-based cancer registries for monitoring cancer burden, which can be assessed in terms of mortality, incidence, health-care use, and economic cost. Planning and evaluating cancer screening programs depend upon disease-stage and incidence data provided by population-based registries. At NIH, the National Cancer Institute has supported the Alaska Native Tumor Registry since 1999 as part of its Surveillance, Epidemiology and End Results (SEER) Program. Data from SEER show that gastric cancer is the third most-common type of cancer in the Alaska Native population (Wiggins 2008). Commonly associated with *Helicobacter pylori* infection, the mortality rate for gastric cancer in the Alaska Native people is more than three times that found in the general U.S. population (Miernyk *et al.* 2011). More research is needed to determine the linkage between *Helicobacter pylori* infection and gastric cancer and peptic ulcer disease in Alaska Natives and other indigenous Arctic populations. In addition, there is an urgent need for effective strategies for treating *Helicobacter pylori* in Arctic populations where infection is endemic. Hepatitis B virus infections occur at endemic rates in Arctic populations and result in high rates of long-term problem such as cirrhosis and cancer of the liver. More needs to be learned, however, about the prevalence and clinical impact of hepatitis C (Young and Bjerregaard 2008).

Native people living in Southwest Alaska suffer a high burden of acute and chronic respiratory disease. For example, one in four infants from the region is hospitalized annually with acute respiratory infections. Hospitalization rates of respiratory syncytial virus (RSV) infection in infants are the highest documented (Karron *et al.* 1999). Bronchiectasis, a chronic lung aftermath of severe pneumonias, is common among Alaska Natives in that region (Singleton *et al.* 2000). A combination of substandard housing, overcrowding, poor indoor-air quality, lack of indoor plumbing and running water, and other environmental factors contribute to such conditions (Hennessy *et al.* 2008). Alaska rates last among U.S. states for adequate sanitation service; approximately 23 percent of rural Alaskan households lack in-home water and sewer service. Hospitalization rates for respiratory, skin, and invasive-bacterial infections are all higher among persons living without running water. Research is needed to reduce environmental triggers of respira-tory disease in homes and develop strategies to address the ongoing disparities in sanitation services and the resultant health impacts.

Problems resulting from trauma and injuries are among the most serious that affect people living and working in the Arctic. Injuries are by far the most important causes of death among people around 35 years of age there (Young and Bjerregaard 2008). While unintentional injuries have always been a hazard of living and working in the Arctic, major sociocultural changes and the widespread availability of alcohol have changed the pattern and extent of these injuries. A thorough review of the underlying cause of these health problems (based upon depression and including alcoholism, other drug abuse, child abuse, and suicide) occurred at a meeting in Anchorage on June 2 and 3, 2009, through collaboration between the USARC and the NIH Fogarty Center. That meeting led to a series of recommendations published in Levintova *et al.*, 2010. Alaska also continues to have a high work-related fatality rate, mostly from drownings in the commercial fishing industry and plane crashes in aviation, which are exacerbated by the cold and remote and Alaska climate. More research is needed to understand the causes and potential interventions needed to reduce the morbidity and mortality associated with intentional (suicide) and unintentional injuries. Special attention should continue to be given to preventing drownings and aircraft crashes. Worker health and safety in oil-spill response in the Arctic or other ice-covered waters poses new concerns as interest in natural resource exploration and extraction increase.

Timeframe

Long-term (5-10 years)

Expected Outcomes

- The circumpolar surveillance of invasive bacterial diseases (causes of pneumonia, meningitis, septicemia) is ongoing and will be used to monitor the impact of intervention programs as they are implemented.
- Together with the Public Health Agency of Canada, ICS will expand to include surveillance of tuberculosis in the U.S. Arctic, northern Canada, Greenland, and six northern regions of the Russian Federation for the collaborative systematic collection, interpretation, and dissemination of information pertaining to tuberculosis in circumpolar populations for use in epidemiologic study, policy generation, program design, and evaluation.
- Use of cancer registry data will allow planning and evaluation of cancer-screening programs and will contribute to the reduction in cancer incidence and mortality.
- Develop consensus guidelines for the diagnosis and treatment of *Helicobacter pylori* in Arctic populations and undertake a circumpolar study on putative bacterial markers of virulence and host and environmental risk factors associated with peptic ulcer disease and gastric cancer in Arctic indigenous peoples. Potential markers will be validated in a prospective study of gastric cancer patients and matched controls to identify persons at risk for early screening.
- Promote and collaborate on surveillance research and management programs of chronic hepatitis B and C using patient registries, increase screening and vaccination for hepatitis B in Arctic, and design collaborative research programs on virology and pathogenesis of hepatitis B and C in the Arctic.
- A North American Free Trade Agreement funded project will measure air quality and implement home-based intervention strategies to reduce levels of home-based environmental triggers of respiratory disease. Additionally, this project will measure the impact of these interventions on the severity and frequency of respiratory symptoms in Alaska Native children with respiratory disease and their families.
- Improve the water and sanitation services available to rural Alaskans by bringing together various stakeholders for the purposes of determining novel strategies, including single home solutions for water purification and sanitation. The goal is to address the ongoing disparities in sanitation services and the resultant health impacts while keeping the costs of solutions manageable.

- Continued development of strategies to prevent morbidity and mortality among workers in Alaska is needed particularly in the commercial fishing and aviation industries.
- A retrospective study of all Arctic maritime disasters and oil spill is being conducted, which will provide detailed information about casualties sustained during those events as well as hazardous exposures suffered by responders, including search and rescue and spill clean-up personnel. The results of this project may be incorporated into U.S. oil spill contingency plans and could be adopted as an addendum to the revised spill response plan being developed by Emergency Prevention, Preparedness, and Response (EPPR) and/or used as source material for occupational safety and health insertions in that plan.
- Develop strategies to prevent morbidity and mortality among oil spill response workers in the Arctic environment. Results may be incorporated into U.S. oil spill contingency plans, and could be adopted as an addendum to the revised spill response plan being developed by EPPR and/ or used as source material for occupational safety and health insertions in that plan.

Milestones

- Publish a comprehensive report on Circumpolar Surveillance of Infectious Diseases for the Arctic Council Ministerial (CDC; FY2013).
- Conduct 5-year retrospective review of tuberculosis in northern Canada, the U.S. Arctic, Greenland and northern regions of the Russian Federation (CDC; FY2013).
- Maintain Alaska's compliance with standards of the National Program of Cancer registries and the National Association of Central Cancer Registries (CDC; FY2012).
- Draft and publish a *Helicobacter pylori* treatment consensus document for high-prevalence. Validate *Helicobacter pylori* bacterial markers of virulence and host and environmental risk factors associated with peptic ulcer disease and gastric cancer in Arctic indigenous peoples (CDC; FY2013).
- Increase screening and vaccination for hepatitis B in Arctic and the design collaborative research programs on virology and pathogenesis of hepatitis B and C in the Arctic (CDC; FY2012).
- Measure indoor air quality before and after home-based intervention strategies in homes of children with chronic respiratory diseases. Measure severity and frequency of respiratory symptoms in children before and after the interventions (CDC, IHS; FY2013).
- Together with the Alaska Department of Environmental Conservation's Village Safe Water Program conduct stakeholder meetings to address scientific, technological, and policy challenges associated with lack of sanitation services in Alaska. Determine health and economic consequences associated with lack of sanitation services in Alaska using epidemiologic studies and economic models (CDC, USARC; FY2013).
- In collaboration with other Arctic nations and the State of Alaska, publish a supplement to the Arctic Council's *Field Guide for Oil Spill Response in Arctic Waters*, which will contribute understanding and control of the unique hazards that exist to workers responding to an oil spill in the Arctic (CDC, NIOSH[13]; FY2013).

- Maintain the Alaska Occupational Injury Surveillance System in cooperation with the State of Alaska. Publish 20 years of occupational safety research to document progress and set goals for the next 10 years (CDC, NIOSH; FY2014).
- Continue work with the USCG and Native Community Development Groups to identify ways to encourage the use of personal flotation devices and personal locator beacons while commercial fishing. Continue partnerships with marine-safety educators to provide cold-water survival classes in remote Native villages (CDC, NIOSH; FY2014).
- Conduct focus groups with pilots and company owners flying to remote villages to identify strategies to combat pilot fatigue, particularly in the high-risk summer months (CDC, NIOSH; FY2013).

Science and Technology Gaps and Needs

Lack of sufficient numbers of certified Tumor Registrars has impeded the technological progress of the Alaska Cancer Registry. The absence of electronic means for hospitals and physicians to report to the central cancer registry is a technological gap. Further technological advances are needed to provide in-home sanitation services in Arctic environments. Challenges include providing sufficient clean water for hand washing and cleanliness and sanitation for the approximately 25% of rural Alaskans without these vital services. Novel solutions for single homes of water provision as well as sanitation need to be explored. Other efforts include improving energy efficiency of water systems operations; alternative delivery and waste removal technologies that address specific problems posed by the climate, soil, and permafrost; and changing source water parameters. Technology should be explored that helps achieves a balance between providing adequate volumes of water to maximize healthy washing behavior and conserving water and energy resources.

3.7.2. Continue Interagency Collaboration to Monitor the Impacts of Climate Change and Environmental Contaminants on Human Health and Wildlife

Why Do This?

Over the last three decades, Arctic Alaska's average temperature has increased by 5 degrees F (2.7 degrees C), resulting in extensive glacier melting, thawing permafrost, reduction in Arctic Ocean sea-ice extent, precipitation increases, decrease in duration of snow cover, longer ice-free seasons on lakes, altered food webs, and greater risk of wildland fires. Warmer temperatures may cause environmental contaminants to move more readily from soil and water into the air; foster greater movement of contaminants from lower-latitude source-area soils (industrial sites, agricultural areas, etc.) into Arctic regions; and lead to contamination of traditional foods with man-made chemicals such as polychlorinated biphenyls (PCBs), dioxins, toxophenes, and other pesticides. In the Arctic, those chemicals may be enriched in the marine- and land-based food webs and end up in humans. These compounds may be associated with adverse health outcomes and have caused some Alaska Natives to decrease consumption of traditional marine subsistence species and substitute with less-healthy western foods (ACIA 2005).

The impact of climate change on communities includes the disruption of permafrost-dependent structures, threats to village water supplies and sanitation systems, reduction in availability of subsistence species, and decreased air quality (wildland fire smoke, dust, and pollens). Health threats from climate change include morbidity and mortality resulting from trauma and injury associated with extreme events (storms, floods, increased heat and cold) and unpredictable ice conditions, increased mental and social stress related to changes in environment and loss of traditional lifestyle, decreased access to quality water and food sources, and potential changes in prevalence of some parasitic and zoonotic infectious diseases (Parkinson 2008). Evidence also exists for health consequences from exposure to toxic metals, such as mercury and lead. There may be adverse effects from exposure to organochlorines and mercury on child neuropsychological development and of organochlorines on their immune function and susceptibility to infection.

New and previously experimental or seasonal shipping lanes are being opened in the Arctic as sea ice retreats and oil and mineral exploration markedly increase. These transport and industrial activities raise the possibility of increasing numbers of shipping disasters and oil spills, with increased risk to human workers, as well as to the marine environment, fauna, and flora. The EPPR Worker Hazard Reduction Project will attempt to address some of those emerging hazards.

Consequently, there is need for studies that collect and analyze human-health and ecosystem observations in rural communities located in the many diverse and ecologically-distinct areas of Alaska. In addition, studies that quantify the cumulative health effects of exposure in Alaska Native mothers and their infants to multiple environmental contaminants [anthropogenic persistent organic pollutants (POPs) and mercury] in subsistence foods are needed.

Timeframe

Mid-term (3-5 years)

Outcomes

- Community health assessments are designed to assess, one village at a time, the existing and emerging threats and to provide residents with the training and capacity to monitor existing threats, detect new threats, and allow the development of adaptation strategies, as proposed in section 3.6. The village-based monitoring programs conducted by ANTHC will form a network that contributes data to state and Federal agencies to better inform those programs responsible for both human and wildlife health. The outcome of these programs could link with developing programs in Canadian Arctic communities.
- Engaging Alaskans as lay observers of human health and ecological events potentially associated with climate change can provide the public-health community with an important early warning of public-health consequences from such change (in conjunction with efforts in 3.6). A University of Alaska, Anchorage project will provide public-health professionals and other decision-makers with a network for the dissemination of informed and culturally appropriate risk communications to guide adaptation planning that reduces health impacts on the Alaskan population.
- A data base of organochlorines, toxic metals, and micronutrients levels in human and salmon tissue will allow the detection of any associations between prenatal exposure to organochlorines and/or mercury with adverse health outcomes, as well as any positive associations between micronutrients levels and health. A time series in human- and subsistence-species tissue levels of organochlorines and mercury will

help evaluate the impact of climate regime change on ocean and atmospheric current delivery of those toxins from lower latitudes into the Bering Sea, addressing possible climate-change impact on subsistence resources and human health. ANTHC will provide and communicate risk and benefit data to Alaska Natives to enable region-wide strategies and policies to reduce the risk and increase benefits of the traditional diet, which will strengthen that critical component of Alaska Native culture.

- Evaluate climate-sensitive infectious-disease surveillance systems for—and baseline levels of infection in—humans and wildlife in Alaska. Such evaluations will result in better understanding of the epidemiology and risk to subsistence wildlife species and to the people who depend of them for food, allowing for rapid detection of outbreaks and development of prevention and control strategies.

Milestones

- Conduct community health assessments and initiate training and deployment of monitoring technology. Develop a web-based monitoring network and village adaptation strategies, and establish the monitoring database. Conduct statistical analysis of laboratory specimens and share data with agencies and jurisdictions. Provide feedback to tribal leaders. Continue planning with interested agencies to develop support for ongoing monitoring and extension to interested communities in Alaska. Disseminate program results to other circumpolar communities and affected communities in the lower 48 states (IHS; FY2015).
- Develop, deploy, and assess a surveillance and response Toolkit for Alaska to promote community based adaptation planning for climate change. Deliverables will include training protocols, a web portal, a surveillance dataset organized by region, adaptation and mitigation recommendations, and reports/manuscripts for peer-reviewed publication (CDC; FY2013).
- Recruit a cohort of 200 Alaska Native women for collection of blood samples, patient interview, and medical-chart information. Collect and chemically analyze salmon tissue samples (CDC, DOI, EPA; FY2015).
- Conduct surveillance evaluations and sero-prevalence studies on humans and wildlife for potentially climate-sensitive infectious diseases such as those caused by *brucella, trichinella, echinococcus, toxoplasma, francisella, giardia,* and *cryptospordium* species (CDC, DOI; FY2013).
- Implement a NOAA-CDC memorandum of agreement for environmental and public health impacts providing exchange of scientific expertise and resources in the areas of climate, weather, water, and environmental, oceanographic, and atmospheric health as it relates to public health (CDC, DOC; FY2015).

Science and Technology Gaps and Needs

Findings from community assessments will require engineering solutions to be developed and applied by the ANTHC Division of Environmental Health and Engineering. Technology will need to be developed for ongoing village-based surveillance and determining antibody levels of zoonotic infectious diseases in killed subsistence animals, as well as village-operated sampling for climate-sensitive microbial threats to water security, such as *giardia, cryptosporidium, toxoplasma, tularemia,* and harmful algal blooms. Development of filter-paper blood-spot tests of animal blood are needed to monitor zoonotic diseases and possible contaminants. There is a general need to improve laboratory diagnostics and molecular-typing systems for many potentially climate-sensitive parasitic and zoonotic infectious agents in humans in wildlife. Finally, adaptation and mitigation planning may require local, state, and national coordination to conduct and evaluate further surveillance measures.

3.7.3. Continue to Support Investigator-Initiated Research in Major Health Priority Areas Such as Mental Health Including Substance Abuse and Suicide, Obesity, Diabetes, and Cancer

Why Do This?

NIH, the leading U.S. biomedical- and behavioral-research agency, supported more than 80 research projects in the Arctic from 2009-2012, with approximate average annual expenditure of $31 million. These projects are conducted by investigators at individual institutions with many having multiple research partners. NIH funding is awarded primarily to research and related institutions in Alaska and the Arctic, to non-Alaska institutions focused on indigenous populations, and to increase biomedical research capacity among Native and non-Native Alaska populations.

The University of Alaska Center for Biomedical Research Excellence (COBRE) program at the Center for Alaska Native Health Research (CANHR) and the *Alaska Institutional Development Award for Biomedical Research Excellence* (INBRE) program receive the largest proportion of NIH funds in the region. Both programs are funded through the National Institute for General Medical Studies to support studies of chemical agents (especially contaminants in subsistence foods) and zoonotic and vector-borne microbial agents of disease. CANHR projects represent one example of multi-agency involvement that includes tribal and local communities, CDC, IHS, and other organizations.

NIH-supported projects in the Arctic also focus on behavioral and mental-health problems, including addiction and related disorders. The National Institute on Minority Health and Health Disparities (NIMHD) and the National Institute on Drug Abuse have supported dozens of research projects on the determinants of substance abuse and mental health disorders in the Arctic. Additionally, NCMHD supports studies of various interventions for substance abuse and mental illness in Alaska.

In addition, NIH has engaged with USARC, CDC, the State of Alaska, and local and academic institutions to develop priorities for research increases in Alaska, particularly on health-care delivery innovations, including telemedicine and information technology. The results of a workshop convened by those entities in Alaska indicate there is major need for better communications among agencies in and out of Alaska on issues related to health research, and there is need for inclusion of local communities or tribal groups in research planning and implementation. Inclusion of innovative and culturally appropriate methods of study is also needed (Levintova *et al.* 2010).

Timeframe

Long-term (5-10 years) Outcomes:

- NIH will continue to support investigator-initiated, peer-reviewed research projects in biomedical and behavioral sciences in the Arctic. Many NIH-awarded projects engage multiple partners.

Science and Technology Gaps and Needs

As need for further funding of biomedical and behavioral research in the Arctic continues, NIH will support appropriate investigator-initiated, peer-reviewed projects. A number of gaps, however, should be highlighted:

- Alaska-based institutions and researchers have not historically been highly competitive peer review. In this time of reduced economic constraints, this gap is difficult to fill. Therefore, there is a substantial need to develop collaborations with successful NIH award recipients.
- It is difficult to recruit from the sparsely populated Native villages in Alaska the large numbers of participants needed to conduct longitudinal studies of chronic diseases. Genetic and cultural similarities among Arctic native populations would allow for statistically meaningful cohorts if communication and information exchange among researchers were optimal.
- Finally, in addition to large population-based studies, exploration of alternative research design (such as small sample design and studies in low-resource settings) is needed.

3.7.4. Continue to Engage Indigenous Communities and Tribal Groups in Research Activities and Projects in the Arctic

Why Do This?

Indigenous community members and their leaders should be involved in all stages of the research process—from formulating and approving projects and methods, to determining research outcomes, to interpreting and disseminating results. Genuine collaboration between researchers and indigenous communities not only builds valuable partnerships and mutual trust, but it also enhances research through information and knowledge sharing. Optimal engagement and partnership efforts enable research that is both culturally sensitive and conducive to producing shared benefits.

Timeframe

Long-term (5-10 years)

Outcomes

- NIH, NSF, and CDC-supported investigators, and others, interested in conducting research in the Arctic/Alaska work closely with community and tribal organizations/leaders in developing and conducting their research. In many cases, projects require tribal permission to be carried out in the community.
- National Heart Lung and Blood Institute provided funding to the Norton Sound Health Corporation, an Alaska Native-owned health corporation, as one of three grantees to conduct the second phase of the Genetics of Coronary Artery Disease in Alaska Natives study.

Milestones

- NIH and CDC supports research through CANHR where researchers collaborate with tribal and local communities on a number research topics, including nutrition, alcohol consumption and abuse prevention, drug abuse and prevention, and others (NIH; FY2012).

Science and Technology Gaps and Needs

Better communication and more information exchanges are necessary between the researchers and the community before the start of research, during, and after its completion.

References

ACIA. (2008). Arctic Climate Impact Assessment Report. Cambridge: Cambridge University Press UK. p863-906.

ACIA. (2005). Arctic Climate Impact Assessment: Scientific Report. 1042 pp. Cambridge University Press, UK.

Bengtsson et al., 2005. Second International Conference on Arctic Research Planning, Science Plan 9, Copenhagen, Denmark. http://www.icarp.dk.

Federal Actions for a Climate Resilient Nation: Progress Report of the Interagency Climate Change Task Force. October 28, 2011. http://www.whitehouse.gov/sites /default/files/microsites/ceq/2011 adapta-tion progress report.pdf.

Hennessy, T.W., Ritter, T., Holman, R., Bruden, D., Yorita, K., Bulkow, L.R., *et al.* (2008). The relationship between in-home water service and the risk of respiratory tract, skin and gastrointestinal tract infections among rural Alaska Natives. *American Journal of Public Health*, 98(11): 2072-2078.

Karron, R., Singleton, R., Bulkow, L., *et al.* (1999). Severe respiratory syncytial virus disease in Alaska Native children. *J Infect Dis* :180, 41-9.

Levintova, M., Zapol, W.I., Engmann, N. (Eds.). (2010). Behavioral and Mental Health Research in the Arctic: A Strategy Setting Meeting. *Inter J. Circumpolar health Suppl. 5*. http://www.ijch.fi/CHS/ CHS 2010(5).pdf.

Miernyk, K., Morris, J., Bruden, D., McMohn, B., Hurlburt, D., Sacco, F., Parkinson, A.J., Hennessy, T., & Bruce, M. (2011). Characterization of helicobacter pylori cagA and vacA genotypes among Alaskans and their correlation with clinical disease. *J.Clin.Micro* 49(9), 3114-3121.

Parkinson, A. J. (2010). Improving human health in the Arctic: the expanding role of the Arctic Council's sustainable development working group. *International J. Circumpolar Health* 69(3), 304-313.

Parkinson, A. J., Bruce, M. G., & Zulz, T. (2008). International Circumpolar Surveillance, and Arctic network for the surveillance of infectious diseases. *Emerg. Infect. Dis.* 14(1), 18-24.

Parkinson, A. J. (2008). Climate Change and Infectious Diseases: the Arctic Environment. IOM Institute of Medicine. Global climate change and extreme weather events: understanding the contributions to infectious disease emergence. Washington DC: The National Academies Press.

Roberts, A. and coauthors (2010). A Science Plan for Regional Arctic System Modeling, A report to the National Science Foundation from the International Arctic Science Community. International Arctic Research Center Technical Papers 10-0001. International Arctic Research Center, University of Alaska Fairbanks.

Wiggins, C. L., Predue, D. G., Henderson, J. A., Bruce, M., Lanier, A. P., Kelly, J. J., Seals, B. F., & Espey, D. K. (2008). Gastric Cancer among American Indians and Alaska Natives in the United States 1999-2004. *Cancer*, 113 (5), 1225-1233.

Singleton, R., *et al.* (2000). Bronchiectasis in Alaska Native children: causes and clinical courses. *Pediatric Pulmonology*, 29(3): 182-187.

Young, K. T, & Bjerregaard, P. (2008). *Health Transitions in Arctic populations*. Toronto: *University of Toronto* Press.

4. Research Infrastructure

Because the Arctic is geographically remote and environmentally harsh, advancing regional knowledge and understanding requires specialized research platforms and instruments. Needed infrastructure ranges from direct on-the-ground observations to satellite observations from space with advanced instruments and from field stations to research vessels. National and international assets are regularly brought to bear. United States infrastructure, its use, and availability are summarized in Tables 1-3. A few relevant foreign assets and missions are also mentioned, but the list is not exhaustive. International coordination of infrastructure and cost sharing is highly desirable.

Table 1 summarizes major infrastructure, including space- and ocean based assets, aircraft (piloted and unpiloted), and field stations. Space-based instruments operated by NASA, NOAA, and other agencies are especially powerful tools for observing the remote Arctic. Reliance on foreign satellites for some observations and measurements is now common. NASA and its partners, including some commercial entities, operate nearly 60 research aircraft, many of which apply to Arctic work. The aircraft most critical to this research plan are listed in Table 1. Research in the Arctic Ocean and adjacent seas employs ice-breaking or ice-strengthened vessels, which are operated by the Coast Guard, the University-National Oceanographic Laboratory System (UNOLS), NOAA, private contractors, and foreign governments. U.S. Navy submarines that deploy occasionally to the Arctic Ocean obtain a limited amount of research-quality data. Sea-ice camps are occasionally operated for brief periods in the Arctic Ocean for process studies and autonomous instrument deployment. As sea-ice cover in the Arctic Ocean diminishes and vessel traffic increases, national needs for Arctic vessels for research, national security, and marine safety are being assessed. Field stations on land and the Greenland Ice Sheet are especially useful for long-term observations and for supporting process studies and shorter-term, satellite field camps.

Table 2 summarizes significant but smaller research tools, with a focus on airborne and ocean-based infrastructure. The latter include numerous autonomous platforms and instruments that are growing in importance as access to research vessels becomes more challenging and the need for year-round observations grows. Basic equipment to enhance indigenous observations is also included.

Table 3 summarizes key assets of cyberinfrastructure that are important for research and education in the Arctic. Electronic media and distribution systems are critical to data acquisition, transmission, archiving and distribution, and to communicate research results to the greater scientific community and public. As data on the Arctic environmental system has become more voluminous, the need for more sophisticated hardware, software, standards, and sharing agreements has increased.

Table 1. Major U.S. infrastructure (space-based, aircraft, ocean-based, field stations) needed to accomplish the five-year Arctic research plan. For each infrastructure element, its use, availability, and relevant sections of the plan are identified

INFRASTRUCTURE	USE	AVAILABILITY	SECTION
Space-based			
Existing satellite missions critical to Arctic research			
NOAA satellite missions	Weather and key climate variables.	Available through 2017.	3.1-3.4
Defense Meteorological Satellite Program (DMSP)	Mapping sea ice with passive microwave.	Available through 2017.	3.1-3.4
NASA Earth Observing Satellites	Detailed studies of sea ice, clouds, and other Arctic parameters.	Many are past design life.	3.1-3.4
Joint Polar Satellite System (JPSS)	Next-generation weather satellite.	SUOMI-NPP has planned operational life to 2017; other satellites are in planning stages.	3.1-3.4
INFRASTRUCTURE	USE	AVAILABILITY	SECTION
USGS Landsat-5 and -7	Agriculture, geology, forestry, regional planning, mapping, global change research, emergency response and disaster relief, education.	Landsat-5 launched in 1984 and still in operation, but data acquisition limited by an electronics problem. Landsat-7 launched in 1999 and still in operation. Minimum design life of 5 years.	3.1-3.4
SAR (Synthetic Aperture Radar)	Sea ice and glacier geophysics and mapping; Marine transportation support; Oceanography; Mapping—vegetation, geology, topography.	No U.S. SAR instruments available. Foreign SAR data (e.g., RADARSAT, TerraSAR-X, COSMO SkyMed) are available for purchase.	3.1-3.4
Satellites planned for launch by 2017			
USGS/NASA LandSat Data Continuity Mission (LDCM)	Agriculture, geology, forestry, regional planning, mapping, global change research, emergency response and disaster relief, education.	Launch in 2013.	3.1-3.4
NASA Global Precipitation Measurement (GPM)	Measure snowfall and heavy rain.	Launch in 2014; Limited footprint over polar regions.	3.3
NASA/DLR (Germany) Gravity Recovery and Climate Experiment (GRACE) follow-on	Arctic oceanography, changes in ice mass, terrestrial water storage.	Launch in 2017.	3.1-3.4
NASA Soil Moisture Active Passive (SMAP)	Soil moisture, freeze thaw patterns, and potentially sea-ice mapping.	Launch in 2015.	3.1-3.4
NASA ICESat 2	Altimetry over land and sea ice to measure changes in thickness.	Launch in 2016.	3.1-3.4

INFRASTRUCTURE	USE	AVAILABILITY	SECTION
Aircraft			
Piloted aircraft			
P-3 (NASA)	Two long-range survey aircraft that can be outfitted with a variety of instruments. Based at NASA Wallops. Currently used for study of Arctic sea ice and land ice, but has applications to many areas of Arctic science.	Available through 2017.	3.1, 3.4
DC-8 (NASA)	One long-range survey aircraft that can be outfitted with a variety of instruments Based at NASA Dryden. Currently used for study of Arctic sea ice and land ice, but has applications to many areas of Arctic science.	Available until 2014, when the aircraft requires major refurbishment.	3.1, 3.3, 3.4
Gulfstream-V (NSF/NCAR)	One long-range, high-altitude aircraft for study of Chemistry and Climate, Chemical Cycles, Air Quality, Mesoscale Weather, Upper Troposphere and Lower Stratosphere.	Available through 2017.	3.3
LC130 (NSF)	Eight ski-equipped heavy-lift aircraft for snow and ice landings to support field operations in Greenland. Based at Stratton Air Base.	Available through 2017.	Indirectly 3.2, 3.3, 3.4
C130Q (NSF/NCAR)	One long-range, high-altitude aircraft with standard thermodynamic, microphysics and radiation sensors, and ability to carry a wide variety of scientific payloads.	Available through 2017.	3.3
C130 (USCG)	Multiple aircraft supporting Homeland Security missions, maritime/Arctic domain awareness, and testing capabilities of personnel and equipment. Scientists and scientific instruments are also accommodated. Based at Coast Guard Air Station Kodiak.	Available through 2017.	3.1, 3.3, 3.4
Twin Otter	NOAA. Four aircraft in the fleet. In the Arctic, aircraft are used for marine mammal surveys in the Bering, Chukchi and Beaufort seas.	Available through 2017.	3.1
Small aircraft	USFWS (DOI) annually contracts light aircraft for fish surveys.		3.2
Unpiloted aircraft			
Global Hawk (NASA and NOAA)	Multiple aircraft supporting atmospheric measurements from high altitude (65,000 feet). Based at NASA Dryden.	Available through 2017. Two aircraft are operational for research, others could be made available.	3.1, 3.3, 3.4
Ikhana (NASA)	One aircraft being developed for sea ice and oceanographic measurements, including buoy drop capability. Based at NASA Dryden.	In development for availability in 2013; availability beyond 2013 dependent on research use.	3.1, 3.4

Table 1. (Continued)

INFRASTRUCTURE	USE	AVAILABILITY	SECTION
Other unpiloted aerial vehicles	Various medium and small UAS and facilities are being developed for physical, chemical, and biological measurements by DOE, NOAA and NSF. DOE is developing a UAS operations facility at Oliktok (Alaska) site by mid-2012.	Available through 2012.	
Ocean-based			
Icebreakers	Physics, chemistry and biology of ice, ocean and atmosphere; Marine geophysics and bathymetry.	The U.S. Coast Guard has three polar icebreakers: Healy, Polar Star, Polar Sea. Healy remains active with an expected service life to 2030. Polar Star is undergoing an extensive overhaul intended to lengthen its service life to approximately 2020-2021 and is expected to be back in service in FY 2013. .	3.1, 3.3, 3.4
		It is expected that Polar Sea will be decommissioned within the year	
Ice-capable research vessels	Physics, chemistry and biology of ice, ocean and atmosphere; Marine geophysics and bathymetry.	The UNOLS vessel R.V. Sikuliaq is scheduled to be available for research in early 2014.	3.1, 3.3, 3.4
Other U.S. research and survey vessels	Physics, chemistry and biology of ice, ocean and atmosphere; Marine geophysics and bathymetry.	The UNOLS vessel R.V. Marcus G. Langseth supported Arctic Ocean geophysical research in summer 2011, demonstrating potential for future use of non-ice strengthened UNOLS vessels.	3.1, 3.3, 3.4
		NOAA seasonally deploys the Oscar Dyson (fishery survey class) and hydrographic survey vessels, Rainier and Fairweather.	
		USFWS (DOI) R.V. Tiglax operates in Arctic waters each summer.	
Foreign vessels	Physics, chemistry and biology of ice, ocean and atmosphere; Marine geophysics and bathymetry.	NOAA annually charters the R.V. Khromova from Russia to carry out the Russian-American Long-term Census of the Arctic—servicing a Bering Strait mooring array and observing climate impacts on sea ice, ocean conditions, and ecosystems.	3.1, 3.3, 3.4

INFRASTRUCTURE	USE	AVAILABILITY	SECTION
		U.S. researchers regularly work from the Canadian Coast Guard icebreakers CCGS Louis St. Laurent and CCGS Wilfred Laurier.	
		International partners (Korea, China, Japan, Russia, Canada) use their vessels to sample in the Distributed Biological Observatory (DBO).	
Submarines	Physics, chemistry and biology of ice, ocean and atmosphere; Marine geophysics and bathymetry.	Occasional (1-3/year) U.S. Navy Science Accommodation Missions (SAMs). Also occasional (every few years) Royal Navy (UK) submarine missions.	3.1, 3.4, 3.5
Sea ice	Physics, chemistry and biology of ice, ocean and atmosphere; Marine geophysics and bathymetry.	Ephemeral facilities that are occupied typically for periods of weeks to months, and less commonly for 1-2 years.	3.1, 3.3, 3.4
Field Stations			
Toolik Lake, North Slope of Alaska	Primarily investigator-led biology and long-term ecological research.	Operations supported by NSF, research supported by NSF and others, and available through 2017.	3.2, 3.4
Summit, Greenland	Primarily investigator led snow and ice geophysics, atmospheric science, and long-term observing. Also host to an emerging NOAA Baseline Observatory for long-term monitoring of atmospheric constituents.	Operations supported by NSF and NOAA, research supported by NSF and others, and available through 2017.	3.2, 3.3, 3.4
Barrow, Alaska	NOAA Baseline Observatory for long-term monitoring of atmospheric constituents.	Available through 2017.	3.2, 3.3, 3.4
	DOE maintains an Atmospheric Radiation Measurement (ARM) research site at Barrow, and a new site at Oliktok, Alaska, will be operational mid-2013.		
	The Next Generation Ecosystem Experiment is expected to operate at Barrow and on the Seward Peninsula, Alaska, beginning in 2012.		
Tiksi, Russia	Primarily atmospheric science.	NSF maintains the observatory in cooperation with NOAA and the Russian Federal Service for Hydrometeorology and Environmental Monitoring.	3.3, 3.4

Table 2. Significant smaller airborne and ocean-based assets and instruments needed to accomplish the five-year Arctic research plan. For each infrastructure element, its use, availability, and relevant sections of the plan are identified

INFRASTRUCTURE	USE	AVAILABILITY	SECTION
Airborne			
LIDAR for ice altimetry	Sea ice thickness. Land ice surface elevation change. NASA Goddard provides various LIDARs for Arctic work for piloted and unpiloted aircraft.	Available through 2017.	3.1, 3.4
Radar, ice penetrating	Snow thickness over sea ice. Bed maps under land ice. Structure of land ice. NASA and NSF support various institutions that develop and operate snow and ice radars.	Available through 2017.	3.1, 3.4
Ocean-based			
Sea ice-based autonomous observatories (e.g., ice-tethered profilers, flux buoys, ice mass balance buoys, wave buoys)	Physics, chemistry and biology of ice, ocean and atmosphere.	Available, but development of new and improved sensors continues.	3.1, 3.3, 3.4, 3.5
Open-ocean autonomous platforms, (e.g., buoys, wavegliders, wave buoys)	Physics, chemistry and biology of ice, ocean and atmosphere.	Available, but development of new and improved sensors and platforms continues.	3.1, 3.3, 3.4, 3.5
Unmanned underwater vehicles (UUV)	Physics, chemistry and biology of water; sea-ice draft; bathymetry.	Short duration: available. Long duration: not available, but under development.	3.1, 3.4, 3.5
Acoustic communication and navigation	Command and control, and data relay to shore, for UUVs and similar platforms.	Not available, but technology exists.	3.1, 3.4
Moorings	Physics, chemistry and biology of water; sea-ice draft.	Available, but development of new and improved sensors continues.	3.1, 3.4, 3.5
Cabled ocean observatories	Physics, chemistry and biology of near-shore waters and ice, and real-time transmission of data; acoustic communication and navigation nodes.	With one exception in the Canadian Arctic, not available.	3.1, 3.4, 3.5
Other			
Navigation (e.g., GPS) and communication tools (e.g., radios)	Technological innovations to enhance the collection of indigenous observations.	Available, but not widely distributed specifically for research-related observations.	3.1, 3.2, 3.4, 3.6, 3.7

Table 3. Cyberinfrastructure needs for the five-year Arctic research plan. For each element, its use, availability, and relevant sections of the plan are identified

CYBERINFRASTRUCTURE	USE	AVAILABILITY	SECTION
Broadband communications and Internet connectivity	Essential components of cyberinfrastructure for research and education.	Broadband communications and connections to the Internet remain a challenge in large areas of the Arctic, thus placing limits on movement of and access to data and information, and education and outreach opportunities for northern residents.	Applies to all sections.
Virtual research networks	Facilitate online communication, collaboration and scientific discovery among researchers.	The NSF Research Coordination Network (RCN) Program addresses the need for the development of virtual research networks. The DOD-funded Arctic Collaborative Environment (ACE) potentially offers a similar capability, with an emphasis on Web-based visualization, analysis and interpretation of diverse datasets.	Applies to all sections.
Data archiving	Long-term data curation and distribution services.	Available for many data sets, e.g., NASA EOSDIS, NOAA National Data Centers (NGDC, NODC, NCDC), DOI/USGS EROS Center, DOE ARM and NGEE data archives. Arctic-relevant data from global and regional model simulations (e.g., Earth System Grid Federation).	Applies to all sections.
		NSF Division of Arctic Sciences funds the development of Advanced Cooperative Arctic Data and Information System (ACADIS) for research data and Exchange for Local Observations and Knowledge (ELOKA) in the Arctic for traditional knowledge.	
		Inter-operability remains a challenge. Human subjects data and traditional knowledge require special handling due to privacy and data ownership concerns.	
Data standards	Adherence to standards from the moment data are acquired makes possible inter-operability of archives, data discovery,	Data standards are in widespread use, but are not necessarily uniform; different standards are used by different disciplines and research communities.	Applies to all sections.

Table 3. (Continued)

CYBERINFRASTRUCTURE	USE	AVAILABILITY	SECTION
	analysis and integration, and derived product development.	Integrated, multi-disciplinary data standards will be needed for those archives that curate and distribute diverse datasets. As it continues to take on a greater role in responsible data and information management, the scientific community is developing and adopting data standards in consultation with data managers. The NSF Coordination Networks (RCN) program encourages RCNs that will develop community standards for data and meta-data.	
Data preparation and processing	QA/QC, preparation of data documentation, metadata profiles and data files, and placement in archives.	Accomplished with varying degrees of quality and success. There is growing recognition that research data must be properly processed and archived. NSF data policy provides guidance, requirements, standards.	Applies to all sections.
Data agreements and access	Facilitate free and open data sharing and exchange nationally and internationally.	In the United States, data are generally freely and openly available, but delays can occur between acquisition and availability in archives. Internationally, the situation is more variable; some countries charge fees and some do not release data.	Applies to all sections.
Data exfiltration	Timely transfer of data from instruments in the field to people for data processing.	Time varies according to circumstances. There is increasing use of satellite communications, e.g., Iridium, for transfer from remote surface instruments. Not available for instruments below the surface leading to delays in data recovery and subsequent broader availability. Broadband communications, including microwave-enabled broadband, remain a challenge in the Arctic—see also 'Broadband communications and Internet connectivity'.	Applies to all sections.

ACRONYMS

Acronym	Full Name
A-CADIS	Advanced Cooperative Arctic Data Information Service
ACIA	Arctic Climate Impact Assessment
ADIWG	Alaska Data Integration Working Group

AEC	Atomic Energy Commission
AeroCan	AERosol CANada
AERONET	Aerosol Robotic Network
AIP	Arctic Investigations Program
AIRS	Atmospheric Infrared Sensor
AMAP	Arctic Monitoring and Assessment Program working group
ANTHC	Alaska Native Tribal Health Consortium
AOMIP	Arctic Ocean Model Intercomparison Project
AON	Arctic Observing Network
AOOS	Alaska Ocean Observing System
ARCPAC	Aerosol, Radiation, and Cloud Processes Affecting Arctic Climate
ARCTAS	Arctic Research on the Composition of the Troposphere from Aircraft and Satellites
ARCUS	Arctic Research Consortium of the United States
ARM	Atmospheric Radiation Measurement
ARPA	Arctic Research and Policy Act of 1984
BOEM	Bureau of Ocean Energy Management
BSSN	Bering Sea Sub-Network
CADIS	Cooperative Arctic Data and Information Service
CAFF	Conservation of Arctic Flora and Fauna working group
CALIOP	Cloud-Aerosol Lidar with Orthogonal Polarisation
CANHR	Center for Alaska Native Health Research
CBMP	Circumpolar Biodiversity Monitoring Program
CCGS	Canadian Coast Guard Ship
CCIN	Canadian Cryospheric Information Network
CDC	Centers for Disease Control and Prevention
CESM	Community Earth System Model
CLIVAR	World Climate Research Programme's Climate Variability and Predictability Program
COBRE	Center for Biomedical Research Excellence (University of Alaska)
CSA	Canadian Space Agency
DBO	Distributed Biological Observatory
DEM	Digital Elevation Model
DHHS	Department of Health and Human Services
DHS	Department of Homeland Security
DOC	Department of Commerce
DOD	Department of Defense
DOE	Department of Energy
DoEd	Department of Education
DOI	Department of the Interior
DOS	Department of State
DOT	Department of Transportation
EALÁT	Reindeer Herders Vulnerability Network Study
EC	Environment Canada
ELOKA	Exchange for Local Observations and Knowledge
EOSDIS	Earth Observing System Data and Information System

EPA	Environmental Protection Agency
EPPR	Emergency Prevention, Preparedness and Response Working Group
EROS	Earth Resources Observation and Science
ESA	European Space Agency
ESMF	Earth System Modeling Framework
ESMs	Earth System Models
FMI	Finnish Meteorological Institute
FWS	Fish and Wildlife Service
GAO	Government Accountability Office
GCMs	General Circulation Models
GINA	Geographic Information Network (University of Alaska)
GIS	Geographic Information Systems
GPS	Global Positioning System
GRACE	Gravity Recovery and Climate Experiment
GRISO	Greenland Ice Sheet and Ocean working group
IARPC	Interagency Arctic Research Policy Committee
IASOA	International Arctic System for Observing the Atmosphere
ICARP II	Second International Conference on Arctic Research Planning
ICESat	Ice, Cloud and land Elevation Satellite
ICS	International Circumpolar Surveillance
IHS	Indian Health Service
INBRE	Institutional Development Award for Biomedical Research Excellence
INTERACT	International Network for Terrestrial Research and Monitoring in the Arctic
IOM	Institute of Medicine
IPCC	Intergovernmental Panel on Climate Change
IPCC AR4	Intergovernmental Panel on Climate Change Fourth Assessment Report
IPY	International Polar Year
ISDAC	Indirect and Semi-Direct Aerosol Campaign
JPSS-1	Joint Polar Satellite System-1
LCCs	Landscape Conservation Cooperatives
LIDAR	Light Detection And Ranging
LTER	Long Term Ecological Research
MISR	Multi-angle Imaging SpectroRadiometer
MIZOPEX	Marginal Ice Zone Observations and Processes Experiment
MMC	Marine Mammal Commission
MODIS	Moderate Resolution Imaging Spectroradiometer
NASA	National Aeronautics and Space Administration
NCDC	National Climatic Data Center
NCMHD	National Center for Minority Health and Disparities
NEH	National Endowment for the Humanities
NGDC	National Geophysical Data Center
NGEE	Next-Generation Ecosystem Experiments
NGO	Non-governmental Organization

NIH	National Institutes of Health
NIOSH	National Institute for Occupational Safety and Health
NMFS-AFSC	National Marine Fisheries Service—Alaska Fisheries Science Center
NOAA	National Oceanic and Atmospheric Administration
NODC	National Oceanographic Data Center
NPP	National Polar-orbiting Partnership
NPS	National Park Service
NSF	National Science Foundation
NSIDC	National Snow and Ice Data Center
NSSI	North Slope Science Initiative
NWS	National Weather Service
ONR	Office of Naval Research
PAG	Pacific Arctic Group
PCBs	Polychlorinated Biphenyls
PI	Principal Investigator
POPs	Persistent Organic Pollutants
QA/QC	Quality Assurance/Quality Control
RADARSAT	Radar Satellite
RSV	Respiratory Syncytial Virus
RV	Research Vessel
SAON	Sustaining Arctic Observing Networks
SBIR	Small Business Innovation Research
SDWG	Sustainable Development Working Group
SEARCH	Study of Environmental Arctic Change
SEER	Surveillance, Epidemiology and End Results
SI	Smithsonian Institution
SIKU	Sea Ice Knowledge and Use
SLCFs	Short lived Climate Forcers
SMAP	Soil Moisture Active Passive
SS	Sea Surface
SSH	Sea Surface Height
SST	Sea Surface Temperature
SWIPA	Snow, Water, Ice, and Permafrost in the Arctic
TES	Tropospheric Emission Spectrometer
UAS	Unpiloted Aerial Systems
UAV	Unmanned Aerial Vehicle
UNOLS	University-National Oceanographic Laboratory System
USARC	United States Arctic Research Commission
USCG	United States Coast Guard
USGS	United States Geological Survey
UUV	Unmanned Underwater Vehicle
VIIRS	Visible Infrared Imager Radiometer Suite
WCRP	World Climate Research Program
WMO	World Meteorological Organization

End Notes

[1] The Arctic Council consists of the eight Arctic States: Canada, Denmark (including Greenland and the Faroe Islands), Finland, Iceland, Norway, Russia, Sweden, and the United States. Six international organizations representing Arctic Indigenous Peoples have permanent participant status.

[2] As this report was being finalized, a new record minimum was recorded for the extent of Arctic sea ice (http://nsidc.org/arcticseaicenews/).

[3] Throughout this document, agencies are listed alphabetically in milestones, and that order does not imply priority.

[4] While this plan covers the period 2013 to 2017, some activities extend work in progress. Some milestones may have ee r eached in 2012.

[5] United States Coast Guard (Department of Homeland Security)

[6] United States Geological Survey (Department of Interior)

[7] The PAG is an informal association of organizations and scientists from Canada, China, Japan, Korea, Russia, and the United States who coordinate ship-based observations and facilitate shared operations in the Pacific sector of the Arctic and work toward data exchange and synthesis.

[8] National Park Service (Department of Interior)

[9] National Weather Service (NOAA, Department of Commerce)

[10] This, and subsequent references to DOS in this section, indicates DOS's role as the Federal lead in international Arctic Council activities, in particular, DOS's role as head of delegation to the Sustainable Development Working Group (SDWG).

[11] National Endowment for the Humanities

[12] Indian Health Service

[13] National Institute for Occupational Safety and Health (Center for Disease Control)

In: The United States in the Arctic
Editor: Joyce B. Stewart

ISBN: 978-1-62618-546-3

Chapter 2

CHANGES IN THE ARCTIC: BACKGROUND AND ISSUES FOR CONGRESS*

Ronald O'Rourke

SUMMARY

The diminishment of Arctic sea ice has led to increased human activities in the Arctic, and has heightened interest in, and concerns about, the region's future. The United States, by virtue of Alaska, is an Arctic country and has substantial interests in the region. On January 12, 2009, the George W. Bush Administration released a presidential directive, called National Security Presidential Directive 66/Homeland Security Presidential Directive 25 (NSPD 66/HSPD 25), establishing a new U.S. policy for the Arctic region.

Record low extents of Arctic sea ice over the past decade have focused scientific and policy attention on links to global climate change and projected ice-free seasons in the Arctic within decades. These changes have potential consequences for weather in the United States, access to mineral and biological resources in the Arctic, the economies and cultures of peoples in the region, and national security.

The five Arctic coastal states—the United States, Canada, Russia, Norway, and Denmark (of which Greenland is a territory)—are in the process of preparing Arctic territorial claims for submission to the Commission on the Limits of the Continental Shelf. The Russian claim to the enormous underwater Lomonosov Ridge, if accepted, would reportedly grant Russia nearly one-half of the Arctic area. There are also four other unresolved Arctic territorial disputes. The diminishment of Arctic ice could lead in coming years to increased commercial shipping on two trans-Arctic sea routes—the Northern Sea Route and the Northwest Passage. Current international guidelines for ships operating in Arctic waters are being updated. Changes to the Arctic brought about by warming temperatures will likely allow more exploration for oil, gas, and minerals. Warming that causes permafrost to melt could pose challenges to onshore exploration activities. Increased oil and gas exploration and tourism (cruise ships) in the Arctic increase the risk of pollution in the region. Cleaning up oil spills in ice-covered waters will be more difficult than in other areas, primarily because effective strategies have yet

* This is an edited, reformatted and augmented version of Congressional Research Service, Publication No. R41153, dated January 2, 2013.

to be developed. Large commercial fisheries exist in the Arctic. The United States is currently meeting with other countries regarding the management of Arctic fish stocks. Changes in the Arctic could affect threatened and endangered species. Under the Endangered Species Act, the polar bear was listed as threatened on May 15, 2008. Arctic climate change is also expected to affect the economies, health, and cultures of Arctic indigenous peoples. Two of the Coast Guard's three polar icebreakers—*Polar Star* and *Polar Sea*—have exceeded their intended 30-year service lives, and *Polar Sea* is not operational. The possibility of increased sea traffic through Arctic waters raises an issue concerning Arctic search and rescue capabilities. On May 12, 2011, representatives from the member states of the Arctic Council signed an agreement on cooperation on aeronautical and maritime search and rescue in the Arctic.

The Arctic has increasingly become a subject of discussion among political leaders of the nations in the region. Although there is significant international cooperation on Arctic issues, the Arctic is also increasingly being viewed by some observers as a potential emerging security issue. In varying degrees, the Arctic coastal states have indicated a willingness to establish and maintain a military presence in the high north. U.S. military forces, particularly the Navy and Coast Guard, have begun to pay more attention to the region.

INTRODUCTION

The diminishment of Arctic sea ice has led to increased human activities in the Arctic, and has heightened interest in, and concerns about, the region's future. Issues such as Arctic sovereignty claims; commercial shipping through the Arctic; Arctic oil, gas, and mineral exploration; endangered Arctic species; and increased military operations in the Arctic could cause the region in coming years to become an arena of international cooperation, competition, or conflict. The United States, by virtue of Alaska, is an Arctic country and has substantial political, economic, energy, environmental, and other interests in the region. Decisions that Congress, the executive branch, foreign governments, international organizations, and commercial firms make on Arctic-related issues could significantly affect these interests.

BACKGROUND[1]

Definitions of the Arctic

There are multiple definitions of the Arctic that result in differing descriptions of the land and sea areas encompassed by the term. Policy discussions of the Arctic can employ varying definitions of the region, and readers should bear in mind that the definition used in one discussion may differ from that used in another. This CRS report does not rely on any one definition.

Arctic Circle Definition and Resulting Arctic Countries

The most common and basic definition of the Arctic defines the region as the land and sea area north of the Arctic Circle (a circle of latitude at about 66.34° North). For surface locations within this zone, the sun is generally above the horizon for 24 continuous hours at least once per year (at the summer solstice) and below the horizon for 24 continuous hours at least once per year (at the winter solstice). The Arctic Circle definition includes the northernmost third or so of Alaska, as well as the Chukchi Sea, which separates that part of Alaska from Russia, and U.S. territorial and Exclusive Economic Zone (EEZ) waters north of Alaska. It does not include the lower two-thirds or so of Alaska or the Bering Sea, which separates that lower part of the state from Russia. Eight countries have territory north of the Arctic Circle: the United States (Alaska), Canada, Russia, Norway, Denmark (by virtue of Greenland, a member country of the Kingdom of Denmark), Finland, Sweden, and Iceland.[2] These eight countries are often referred to as the Arctic countries, and they are the member states of the Arctic Council, an intergovernmental forum established in 1996.[3] A subset of the eight Arctic countries are the five countries that are considered Arctic coastal states: the United States, Canada, Russia, Norway, and Denmark (by virtue of Greenland).

Definition in Arctic Research and Policy Act (ARPA) of 1984

Section 112 of the Arctic Research and Policy Act (ARPA) of 1984 (Title I of P.L. 98-373 of July 31, 1984)[4] defines the Arctic as follows:

> As used in this title, the term "Arctic" means all United States and foreign territory north of the Arctic Circle and all United States territory north and west of the boundary formed by the Porcupine, Yukon, and Kuskokwim Rivers [in Alaska]; all contiguous seas, including the Arctic Ocean and the Beaufort, Bering, and Chukchi Seas; and the Aleutian chain.

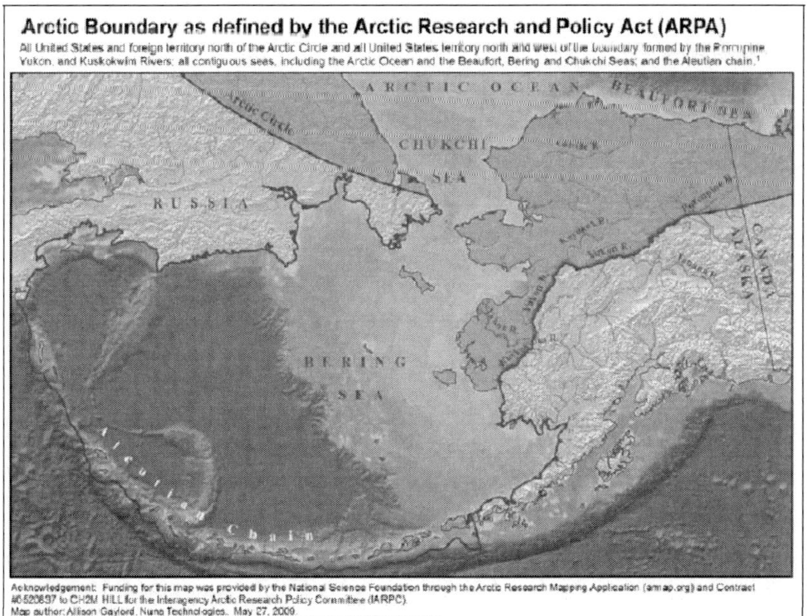

Source: U.S. Arctic Research Commission (http://www.arctic.gov/maps/ARPA_ Alaska_only_ 150dpi.jpg, accessed on December 23, 2011).

Figure 1. Arctic Area of Alaska as Defined by ARPA.

This definition, which is codified at 15 U.S.C. 4111,[5] includes certain parts of Alaska below the Arctic Circle, including the Aleutian Islands and portions of central and western mainland Alaska, such as the Seward Peninsula and the Yukon Delta. Figure 1. below shows the Arctic area of Alaska as defined by ARPA.

Figure 2 shows the entire Arctic area as defined by ARPA.

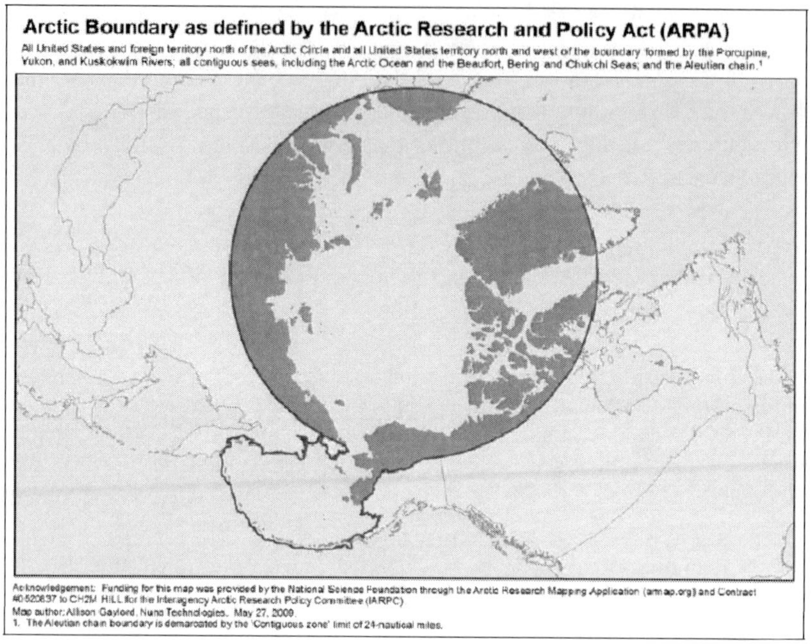

Source: U.S. Arctic Research Commission (http://www.arctic.gov/maps/ARPA_Polar_ 150dpi.jpg, accessed on December 23, 2011).

Figure 2. Entire Arctic Area as Defined by ARPA.

Other Definitions

Other definitions of the Arctic are based on factors such as average temperature, the northern tree line, the extent of permafrost on land, the extent of sea ice on the ocean, or jurisdictional or administrative boundaries.[6] A definition based on a climate-related factor could circumscribe differing areas over time as a result of climate change.

The 10° C isotherm definition of the Arctic defines the region as the land and sea area in the northern hemisphere where the average temperature for the warmest month (July) is below 10° Celsius, or 50° Fahrenheit. This definition results in an irregularly shaped Arctic region that excludes some land and sea areas north of the Arctic Circle but includes some land and sea areas south of the Arctic Circle. This definition currently excludes all of Finland and Sweden, as well as some of Alaska above the Arctic Circle, while including virtually all of the Bering Sea and Alaska's Aleutian Islands.[7]

The definition of the Arctic adopted by the Arctic Monitoring and Assessment Programme (AMAP)—a working group of the Arctic Council—"essentially includes the terrestrial and marine areas north of the Arctic Circle (66°32' N), and north of 62° N in Asia and 60° N in North America, modified to include the marine areas north of the Aleutian chain, Hudson Bay, and parts of the North Atlantic, including the Labrador Sea."[8] The AMAP website includes a map showing the Arctic Circle, 10° C isotherm, tree line, and AMAP definitions of the Arctic.[9]

Some observers use the term "high north" as a way of referring to the Arctic. Some observers make a distinction between the "high Arctic"—meaning, in general, the colder portions of the Arctic that are closer to the North Pole—and other areas of the Arctic that are generally less cold and further away from the North Pole, which are sometimes described as the low Arctic or the subarctic.

U.S. Arctic Research

Arctic Research and Policy Act (ARPA) of 1984, As Amended

The Arctic Research and Policy Act (ARPA) of 1984 (Title I of P.L. 98-373 of July 31, 1984)[10] "provide[s] for a comprehensive national policy dealing with national research needs and objectives in the Arctic."[11] The act, among other things

- made a series of findings concerning the importance of the Arctic and Arctic research;
- established the U.S. Arctic Research Commission (USARC) to promote Arctic research and recommend Arctic research policy;
- designated the National Science Foundation (NSF) as the lead federal agency for implementing Arctic research policy;
- established the Interagency Arctic Research Policy Committee (IARPC) to develop a national Arctic research policy and a five-year plan to implement that policy, and designated the NSF representative on the IARPC as its chairperson;[12] and
- defined the term "Arctic" for purposes of the act.

The Arctic Research and Policy Act of 1984 was amended by P.L. 101-609 of November 16, 1990. For the texts of the Arctic Research and Policy Act of 1984 and P.L. 101-609, see Appendix A and Appendix B, respectively.

FY2013 NSF Budget Request For Arctic Research

NSF—the lead federal agency for implementing Arctic research policy (see "Arctic Research and Policy Act (ARPA) of 1984, As Amended")—carries out Arctic research activities through its Division of Arctic Sciences (ARC), which forms part of NSF's Office of Polar Programs (OPP). NSF is requesting a total of $449.74 million for OPP for FY2012, including $108.51 million for ARC.[13] NSF states in its budget request that OPP

> The Office of Polar Programs (OPP) is the primary U.S. supporter of fundamental research in the polar regions. In addition, NSF provides interagency leadership for U.S. activities in polar regions. In the Arctic, NSF helps coordinate research planning as directed by the Arctic Research Policy Act of 1984. The NSF Director chairs the Interagency Arctic Research Policy Committee created for this purpose, which is now a component of the President's National Science and Technology Council....
>
> OPP's FY 2013 Budget Request is influenced by four key priorities: (1) maintaining strong disciplinary programs that provide a base for our investments in cross-disciplinary

system science programs; (2) strengthening U.S. research community activities in polar system science; (3) supporting critical facilities that enable research in the Earth's polar regions; and (4) enabling scientific leadership in research in polar regions that is aligned with the NSF-wide investments in Science, Engineering, and Education for Sustainability (SEES) and Cyberinfrastructure Framework for 21st Century Science and Engineering (CIF21). These priorities reflect opportunities for fundamental scientific discovery uniquely available in polar regions, as well as studies to investigate the causes and future trajectory of changes now being observed at the poles that could impact global systems. This work will implement the Foundation's lead-agency role in facilitating the Nation's investment in polar science, where environmental change in parts of the Arctic and Antarctic is occurring faster than anywhere else in the world, and has a wide variety of regional and global impacts. Research supported by OPP will elucidate the causes and likely impacts of these changes, thus providing a sound basis for future policy decisions.

In FY 2013 OPP will also provide support for NSF-wide efforts, such as Integrating NSF Support Promoting Interdisciplinary Research and Education (INSPIRE) and NSF Innovation Corps (I-Corps).[14]

Regarding the $108.51 million requested for ARC for FY2013, NSF states that

ARC's FY 2013 Request is focused on fundamental research and the science needed by policymakers to support decision-making regarding sound and sustainable economic and social development. A common theme in these activities is that they support NSF-funded research in the broader context of both interagency and international scientific studies....

Arctic Sciences is organized into several programs that support research in social science, earth systemscience, and a broad range of natural sciences. Educational projects are also supported.

The Research Support and Logistics program assists researchers with access to the Arctic, improves safety and environmental stewardship, and increases the ability of researchers to share plans and results with local Arctic communities. The Arctic is experiencing rapid climate change, with scientific observations recording an estimated 14 percent per decade reduction in sea ice extent in the Arctic over the past 30 years, and significant summer melting of the Greenland Ice Sheet.

At the same time the changes vary from region to region and, in addition, are tightly coupled to climate variability, which is also regional. These and other phenomena are forcing change and uncertainty in traditional Arctic populations, presenting challenges and opportunities for industry and commerce, and have the potential to affect the global population through changes in sea level and changed weather patterns. Arctic Sciences funds a broad range of activities to provide an integrated understanding of environmental change in the Arctic, including the study of significant, system-scale environmental change and its human dimension.

The Research Support and Logistics program is driven by and responds to research and education funded by the division.

Funding is provided directly to grantees or to key organizations that provide or manage Arctic support and logistics. Emphasis will be placed on improving access to and the energy security of the remote facilities used by Arctic researchers and educators. In general, 40 percent of the Arctic Science division's portfolio is available for new research grants, with 20 percent for continuing grants made in previous years and 40 percent for research support and logistics.[15]

January 2009 Arctic Policy Directive (NSPD 66/HSPD 25)

On January 12, 2009, the George W. Bush Administration released a presidential directive establishing a new U.S. policy for the Arctic region. The directive, dated January 9, 2009, was issued as National Security Presidential Directive 66/Homeland Security Presidential Directive 25 (NSPD 66/HSPD 25). The directive was the result of an interagency review, and it superseded for the Arctic (but not the Antarctic) a 1994 presidential directive on Arctic and Antarctic policy. The directive, among other things,

- states that the United States is an Arctic nation, with varied and compelling interests in the region;
- sets forth a six-element overall U.S. policy for the region;
- describes U.S. national security and homeland security interests in the Arctic; and
- discusses a number of issues as they relate to the Arctic, including international governance; the extended continental shelf and boundary issues; promotion of international scientific cooperation; maritime transportation; economic issues, including energy; and environmental protection and conservation of natural resources.

The Obama Administration has not issued a new directive superseding NSPD 66/HSPD 25; it is currently operating under the Bush Administration's policy directive.[16] For the text of NSPD 66/HSPD 25, see Appendix C.

May 2010 National Security Strategy

In May 2010, the Obama Administration released a national security strategy document, which states:

> The United States is an Arctic Nation with broad and fundamental interests in the Arctic region, where we seek to meet our national security needs, protect the environment, responsibly manage resources, account for indigenous communities, support scientific research, and strengthen international cooperation on a wide range of issues.[17]

The Arctic and the U.N. Convention on Law of the Sea (UNCLOS)[18]

Background to UNCLOS

In November 1994, the United Nations Convention on the Law of the Sea (UNCLOS) entered into force. This convention establishes a treaty regime to govern activities on, over, and under the world's oceans. It builds on the four 1958 law of the sea conventions and sets forth a framework for future activities in parts of the oceans that are beyond national jurisdiction.[19] The 1982 Convention and its 1994 Agreement relating to Implementation of Part XI of the Convention were transmitted to the Senate on October 6, 1994.[20] In the absence of Senate advice and consent to adherence, the United States is not a party to the convention and agreement.

Part VI of UNCLOS and Commission on Limits of Continental Shelf

Part VI of the convention, dealing with the Continental Shelf, and Annex II, which established a Commission on the Limits of the Continental Shelf, are most pertinent to the Arctic as it becomes more accessible ocean space, bordered by five coastal states.[21] The convention gives the coastal state sovereign jurisdiction over the resources, including oil and gas, of its continental shelf.[22]

Under Article 76 of the convention, a coastal state with a broad continental margin may establish a shelf limit beyond 200 nautical miles. This jurisdiction is subject to the submission of the particulars of the intended limit and supporting scientific and technical data by the coastal state to the commission for review and recommendation.[23] The commission reviews the documentation and, by a two-thirds majority, approves its recommendations to the submitting state. Coastal states agree to establish the outer limits of their continental shelf, in accordance with this process and with their national laws. In instances of disagreement with the commission's recommendations, the coastal state may make a revised or new submission. The actions of the commission "shall not prejudice matters relating to delimitation of boundaries between States with opposite or adjacent coasts."[24] The "limits established by a coastal State on the basis of these recommendations shall be final and binding."[25]

U.S. Activities as a Non-Party to UNCLOS

As a non-party to the convention, the United States cannot participate as a member of the commission; it cannot submit a claim under Article 76. Over the years, however, it has submitted observations on submissions made by other states, requesting that those observations be made available online and to the commission. In addition, since 2001, the United States has gathered and analyzed data to determine the outer limits of its extended continental shelf. Starting in 2007, this effort became the Extended Continental Shelf Project, directed by an interagency task force under the lead of the Department of State.[26] Some observers have suggested that a separate regime be negotiated to address the changing circumstances in the Arctic. They maintain that this phenomenon was not envisioned at the time the Law of the Sea Convention was negotiated. Still others suggest that the Arctic region above a certain parallel be designated a wilderness area, and they cite as precedent Article 4 of The Antarctic Treaty, under which any current claims to sovereign territory are frozen and

> No acts or activities taking place while the present Treaty is in force shall constitute a basis for asserting, supporting or denying a claim to territorial sovereignty in Antarctica or create any rights of sovereignty in Antarctica. No new claim, or enlargement of an existing claim, to territorial sovereignty in Antarctica shall be asserted while the present Treaty is in force.

Supporters of the Law of the Sea Convention maintain that changing circumstances in the Arctic strengthen their argument that the United States should become a party to the convention. In this way, they argue, the United States can be best situated to protect and serve its national interests, under both Article 76 and other parts of the convention.

ISSUES FOR CONGRESS

Climate Change and Loss of Arctic Sea Ice[27]

Record low extents of Arctic sea ice in 2012 and 2007 have focused scientific and policy attention on climate changes in the High North, and to the implications of projected ice-free[28] seasons in the Arctic within decades. The Arctic has been projected by several scientists to be ice-free in most late summers as soon as the 2030s.[29] This opens opportunities for transport through the Northwest Passage and the Northern Sea Route, extraction of potential oil and gas resources, and expanded fishing and tourism (Figure 3). More broadly, physical changes in the Arctic include warming ocean, soil, and air temperatures; melting permafrost; shifting vegetation and animal abundances; and altered characteristics of Arctic cyclones. All these changes are expected to affect traditional livelihoods and cultures in the region, survival of polar bear and other animal populations, and raise risks of pollution, food supply, safety, cultural losses, and national security. Moreover, linkages ("teleconnections") between warming Arctic conditions and extreme events in the mid-latitude continents are increasingly evident, identified in such extreme events as the heat waves and fires in Russia in 2010; severe winters in the eastern United States and Europe in 2009/2010 and in Europe in 2011/2012,[30] and Indian summer monsoons and droughts. Hence, changing climate in the Arctic suggests important implications both locally and across the Hemisphere. Like the rest of the globe, temperatures in the Arctic have varied[31] but show a significant warming trend since the 1970s, and particularly since 1995.[32] The annual average temperature for the Arctic region (from 60° to 90° N) is now about 1.8° F warmer than the "climate normal" (the average from 1961 to 1990). Temperatures in October-November are now about 9° F above the seasonal normal. Scientists have concluded that most of the *global* warming of the last three decades is very likely caused by human-related emissions of greenhouse gases (GHG, mostly carbon dioxide); they expect the GHG-induced warming to continue for decades, even if, and after, GHG concentrations in the atmosphere have been stabilized. The extra heat in the Arctic is amplified by processes there (the "polar amplification") and may result in irreversible changes on human timescales. The observed warmer temperatures along with rising cyclone size and strength in the Arctic have reduced sea ice extent, thickness, and ice that persists year-round ("perennial ice"); natural climate variability has likely contributed to the record low ice extents of 2007 and 2012. The 2007 minimum sea ice extent was influenced by warm Arctic temperatures and warm, moist winds blowing from the North Pacific into the central Arctic, contributing to melting and pushing ice toward and into the Atlantic past Greenland. Warm winds did not account for the near-record sea ice minimum in 2008.[33] In early August 2012, an unusually large storm with low pressure developed over the Arctic, helping to disperse the already weak ice into warmer waters and accelerating its melt rate. By August 24, 2012, sea ice extent had shrunk below the previous observed minimum of late September 2007.[34] Modeling of GHG-induced climate change is particularly challenging for the Arctic, but it consistently projects warming through the 21st century, with annual average Arctic temperature increases ranging from +1° to +9.0° C (+2° to +19.0° F), depending on the GHG scenario and model used. While such warming is projected by most models throughout the Arctic, some models project slight cooling localized in the North Atlantic Ocean just south of Greenland and Iceland. Most warming would occur in autumn and winter, "with very little temperature change projected over the Arctic Ocean" in summer months.[35] Due to observed and projected climate change, scientists have concluded that the Arctic will have changed from an ice-covered environment to a recurrent ice-free[36] ocean (in summers) as soon as the late 2030s. The character of ice cover is expected to change as well, with the ice being thinner, more fragile, and more regionally variable. The variability in recent years of both ice quantity and location could be expected to continue.

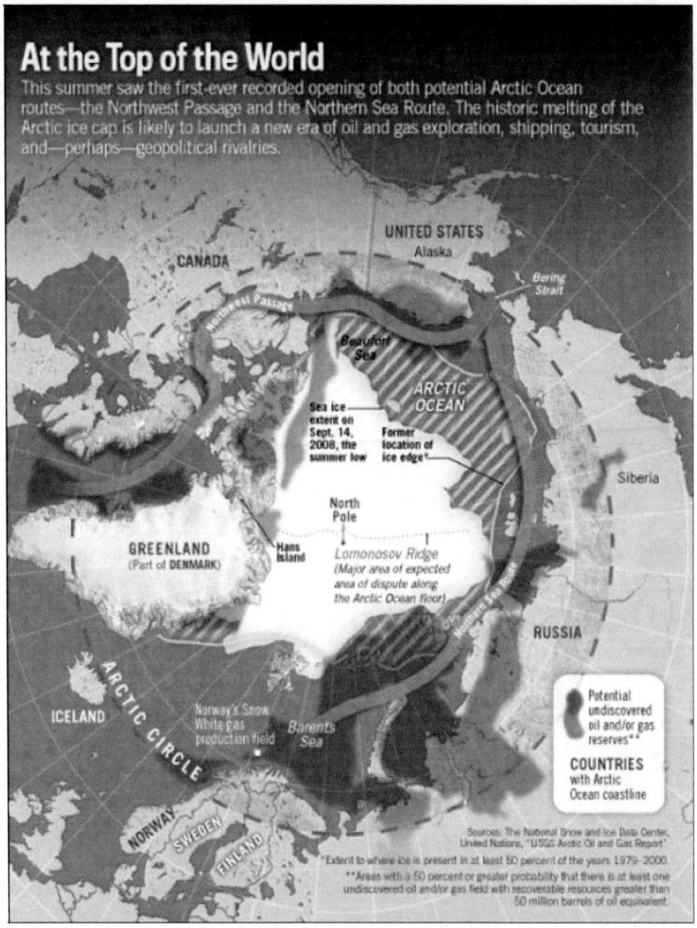

Source: Graphic by Stephen Rountree at *U.S. News and World Report*, http://www.usnews.com/articles /news/world/2008/10/09/global-warming-triggers-an-international-race-for-the-artic/photos/#1.

Figure 3. Arctic Sea Ice Extent in September 2008, Compared with Prospective Shipping Routes and Oil and Gas Resources.

Territorial Claims and Sovereignty Issues[37]

Motivated in part by a desire to exercise sovereign control over the Arctic region's increasingly accessible oil and gas reserves (see "Oil, Gas, and Mineral Exploration"), the four Arctic coastal states other than the United States—Canada, Russia, Norway, and

Denmark (of which Greenland is a territory)—are in the process of preparing territorial claims in the Arctic, including claims for expanded Exclusive Economic Zones (EEZs), for submission to the Commission on the Limits of the Continental Shelf. (As mentioned earlier—see "U.S. Activities As a Non-Party to UNCLOS"—the United States, as a non-party to UNCLOS, cannot participate as a member of the commission; it cannot submit a claim under Article 76. Over the years, however, it has submitted observations on submissions made by other states, requesting that those observations be made available online and to the commission. In addition, since 2001, the United States has gathered and analyzed data to determine the outer limits of its extended continental shelf.)

Russia has been attempting to chart the Arctic Ocean's enormous underwater Lomonosov Ridge in an attempt to show that it is an extension of Russia's continental margin. The Russian claim to this ridge, if accepted, would reportedly grant Russia nearly one-half of the Arctic area; a 2001 claim submitted by Russia was rejected as insufficiently documented. Canada also claims a portion of the Lomonosov Ridge as part of its own underwater continental shelf.[38] In August 2007, a Russian submersible on a research expedition deposited an encased Russian Federation flag on the seabed of the presumed site of the North Pole. The action captured worldwide attention, but analysts note that it did not constitute an official claim to the territory and was therefore a purely symbolic act.

At a May 2008 meeting in Ilulissat, Greenland, the five Arctic coastal states reaffirmed their commitment to the UNCLOS legal framework for the establishment of extended continental shelf limits in the Arctic.[39] (For further discussion, see "Extent of the Continental Margin" in "Oil, Gas, and Mineral Exploration.")

In addition to this process, there are four unresolved Arctic territorial disputes:

- Scientists have forecast that in coming decades, global warming will reduce the ice pack in Canada's northern archipelago sufficiently to permit ships to use the trans-Arctic shipping route known as the Northwest Passage during the summer months (see "Commercial Sea Transportation"). The prospect of such traffic raises a major jurisdictional question. Ottawa maintains that such a passage would be an inland waterway, and would therefore be sovereign Canadian territory subject to Ottawa's surveillance, regulation, and control. The United States, the European Union, and others assert that the passage would constitute an international strait between two high seas.
- The United States and Canada are negotiating over a binational boundary in the Beaufort Sea.
- The United States and Russia in 1990 signed an agreement regarding a disputed area of the Bering Sea; the U.S. Senate ratified the pact the following year, but the Russian *Duma* has yet to approve the accord.
- Denmark and Canada disagree over which country has the territorial right to Hans Island, a tiny, barren piece of rock between Greenland and Canada's Ellesmere Island. Some analysts believe the two countries are vying for control over a future sea lane that might be created if the Arctic ice were to melt sufficiently to create a Northwest Passage. Others claim that the governments are staking out territorial claims in the event that future natural resource discoveries make the region economically valuable.[40]

In addition to these disputes, Norway and Russia had been at odds for decades over the boundary between the two in the so-called "Grey Zone" in the Barents Sea, an area believed to hold rich undersea deposits of petroleum. On September 15, 2010, Norwegian Prime Minister Jens Stoltenberg and Russian President Dmitry Medvedev signed an agreement in Murmansk, a Russian city near the Norwegian border. The accord awards roughly half of the 175,000-squarekilometer area to each country; it spells out fishing rights, and provides for the joint development of future oil and gas finds that straddle the boundary line. Some observers believe it is noteworthy that Russia would concede sovereignty over such a large, resource-rich territory to a small, neighboring country. But others have noted that Moscow may be hoping for Norwegian cooperation in developing offshore resources, and eventually in winning approval when Russia submits its Article 76 UNCLOS claim.[41]

In August 2010, Canadian Foreign Minister Lawrence Cannon announced a new "Statement of Canada's Arctic Policy," which reaffirmed the government's commitment to Canada's sovereignty in the region, to economic and social development, to environmental protection, and to empowerment of the peoples in the north. The statement also emphasized the government's intention to negotiate settlements to its disputes with the United States over the Beaufort Sea boundary, and with Denmark over Hans Island. Minister Cannon declared that "making progress on outstanding boundary issues will be a top priority."[42] Also, despite their dispute over Hans Island, Canada and Denmark have been working together on Arctic issues. In May 2010, the two countries' military chiefs of staffs signed a memorandum of understanding on Arctic Defense, Security, and Operational Cooperation, committing the two countries to "enhanced consultation, information exchange, visits, and exercises."[43]

Commercial Sea Transportation[44]

Background

The search for a shorter route from the Atlantic to Asia has been the quest of maritime powers since the Middle Ages. The melting of Arctic ice raises the possibility of saving several thousands of miles and several days of sailing between major trading blocs.[45] If the Arctic were to become a viable shipping route, the ramifications could extend far beyond the Arctic. For example, lower shipping costs could be advantageous for China (at least its northeast region), Japan, and South Korea because their manufactured products exported to Europe or North America could become less expensive relative to other emerging manufacturing centers in Southeast Asia, such as India.[46] Melting ice could potentially open up two trans-Arctic routes (see Figure 3):[47]

- **The Northern Sea Route** (NSR, a.k.a. the "Northeast Passage"), along Russia's northern border from Murmansk to Provideniya, is about 2,600 nautical miles in length. It was opened by the Soviet Union to domestic shipping in 1931 and to transit by foreign vessels in 1991. Even so, these waters are little used by non-Russian ships.[48] This route would be applicable for trade between northeast Asia (north of Singapore) and northern Europe.
- **The Northwest Passage** (NWP) runs through the Canadian Arctic Islands. The NWP actually consists of several potential routes. The southern route is through Peel Sound in Nunavut, which has been open in recent summers and contains mostly one-

year ice. However, this route is circuitous, contains some narrow channels, and is shallow enough to impose draft restrictions on ships. The more northern route, through McClure Strait from Baffin Bay to the Beaufort Sea north of Alaska, is much more direct, and therefore more appealing to ocean carriers, but more prone to ice blockage.[49] The NWP is potentially applicable for trade between northeast Asia (north of Shanghai) and the northeast of North America, but may be less commercially viable than the NSR.[50]

Destination Traffic, Not Trans-Arctic Traffic

Almost all cargo ship activity currently taking place in the Arctic is to transport natural resources from the Arctic or to deliver general cargo and supplies to communities and natural resource extraction facilities. Thus, cargo ship traffic in the Arctic presently is mostly regional, not trans-Arctic. While there has been a recent uptick in Arctic shipping activity, this activity has more to do with a spike in commodity prices than it does with the melting of Arctic ice. Even so, recent activity is less than it has been in the past. The NSR continues to account for the bulk of Arctic shipping activity. The western end of the NSR, in the Barents Sea, is open year round. Further east, sections of the NSR are open for as little as two and a half months during the summer. However, in 2011, the route was open for five months and 34 ships, accompanied by a Russian icebreaker, passed through.[51]

Cruise Ship Activity

Considerable cruise ship activity takes place in Arctic waters. In the summer of 2007, three cruise ships reportedly sailed through the NWP from the Atlantic to Alaska's North Slope.[52] In August 2010, a cruise ship with over 150 passengers ran aground in the NWP. In the Barents Sea, there are regular cruise ships sailing to Svalbard. The inherent dangers for passenger ships in the Arctic have prompted calls for international regulations promoting the safety of cruise ships in the area. Some have suggested that cruise ships sail in pairs to provide assistance to one another, given the Arctic's remoteness and the difficulty land-based rescuers would have in reaching a vessel in distress.[53] Requiring that Arctic cruise vessels have ice-strengthened hulls and be equipped with enclosed lifeboats could be other safety requirements. In 2003, some Arctic cruise and tourist operators formed the Association of Arctic Expedition Cruise Operators (AECO) to establish agreed-upon safety and environmental protection guidelines, but this organization only covers the portion of the Arctic around Greenland, Svalbard, and Jan Mayen.[54]

Unpredictable Ice Conditions Hinder Trans-Arctic Shipping

Arctic waters do not necessarily have to be ice free to be open to shipping. Multiyear ice can be over 10 feet thick and problematic even for icebreakers, but one-year ice is typically 3 feet thick or less. This thinner ice can be more readily broken up by icebreakers or ice class ships (cargo ships with reinforced hulls and other features for navigating in ice-infested waters). However, more open water in the Arctic has resulted in another potential obstacle to shipping: unpredictable ice flows. In the NWP, melting ice and the opening of waters that were once covered with one-year ice has allowed blocks of multiyear ice from farther north, or icebergs from Greenland, to flow into potential sea lanes. The source of this multiyear ice is not predicted to dissipate in spite of climate change. Moreover, the flow patterns of these ice blocks are very difficult to predict, and they have floated into potential routes for shipping.[55] Thus, the lack of ice in potential sea lanes during the summer months can add even greater unpredictability to Arctic shipping. This is in addition to the extent of ice versus open water, which is also highly variable from one year to the next and seasonally.

The unpredictability of ice conditions is a major hindrance for trans-Arctic shipping in general, but can be more of a concern for some types of ships than it is for others. For instance, it would be less of a concern for cruise ships, which may have the objective of merely visiting the Arctic rather than passing through and could change their route and itinerary depending on ice conditions. On the other hand, unpredictability is of the utmost concern for container ships that carry thousands of containers from hundreds of different customers, all of whom expect to unload or load their cargo upon the ship's arrival at various ports as indicated on the ship's advertised schedule. The presence of even small blocks of ice or icebergs from a melting Greenland ice sheet requires slow sailing and could play havoc with schedules. Ships carrying a single commodity in bulk from one port to another for just one customer have more flexibility in terms of delivery windows, but would not likely risk an Arctic passage under prevailing conditions.

Ice is not the sole impediment to Arctic shipping. The region frequently experiences adverse weather, including not only severe storms, but also intense cold, which can impair deck machinery. During the summer months when sea lanes are open, heavy fog is common in the Arctic.

Commercial ships would face higher operating costs on Arctic routes than elsewhere. Ship size is an important factor in reducing freight costs. Many ships currently used in other waters would require two icebreakers to break a path wide enough for them to sail through; ship owners could reduce that cost by using smaller vessels in the Arctic, but this would raise the cost per container or per ton of freight.[56] Also, icebreakers or ice-class cargo vessels burn more fuel than ships designed for more temperate waters and would have to sail at slower speeds. The shipping season in the Arctic only lasts for a few weeks, so icebreakers and other special required equipment would sit idle the remainder of the year. None of these impediments by themselves may be enough to discourage Arctic passage but they do raise costs, perhaps enough to negate the savings of a shorter route. Thus, from the perspective of a shipper or a ship owner, shorter via the Arctic does not necessarily mean cheaper and faster.[57]

Basic Navigation Infrastructure Is Lacking

Considerable investment in navigation-related infrastructure would be required if trans-Arctic shipping were to become a reality. Channel marking buoys and other floating visual aids are not possible in Arctic waters because moving ice sheets will continuously shift their positions. Therefore, vessel captains would need to rely on marine surveys and ice charts. For some areas in the Arctic, however, these surveys and charts are out of date and not sufficiently accurate.[58] To remedy this problem, aviation reconnaissance of ice conditions and satellite images would need to become readily available for ship operators.[59] Ship-to-shore communication infrastructure would need to be installed where possible. Refueling stations may be needed, as well as, perhaps, transshipment ports where cargo could be transferred to and from ice-capable vessels at both ends of Arctic routes. Shipping lines would need to develop a larger pool of mariners with ice navigation experience. Marine insurers would need to calculate the proper level of risk premium for polar routes, which would require more detailed information about Arctic accidents and incidents in the past.

The U.S. Army Corps of Engineers, along with the state of Alaska, is studying the feasibility of a "deep-draft" port in the Arctic (accommodating ships with a draft of up to 35 feet).

The northern and northwestern coastlines of Alaska are exceptionally shallow, generally limiting harbor and near shore traffic to shallow-draft barges. Coast Guard cutters and icebreakers have drafts of 35 to 40 feet while NOAA research vessels have drafts of 16 to 28 feet, so at present these vessels are based outside the Arctic and must sail considerable distances to reach Arctic duty stations. Supply vessels supporting offshore oil rigs typically have drafts over 20 feet. A deep-draft port could serve as a base of operations for larger vessels, facilitating commercial maritime traffic in the Arctic.[60]

Another study on maritime infrastructure needs is being conducted by the U.S. Committee on the Marine Transportation System, a Cabinet-level committee of federal agencies with responsibilities for marine transportation. It is due to report to Congress in 2012 a priority of federal actions the committee believes are needed to ensure maritime safety, environmental protection, and national security in the Arctic.[61]

Regulation of Arctic Shipping

Due to the international nature of the shipping industry, maritime trading nations have adopted international treaties that establish standards for ocean carriers in terms of safety, pollution prevention, and security. These standards are agreed upon by shipping nations through the International Maritime Organization (IMO), a United Nations agency that first met in 1959.[62]

Key conventions that the 168 IMO member nations have adopted include the Safety of Life at Sea Convention (SOLAS), which was originally adopted in response to the *Titanic* disaster in 1912 but has since been revised several times; the Prevention of Pollution from Ships (MARPOL), which was adopted in 1973 and modified in 1978; and the Standards for Training, Certification, and Watchkeeping for Seafarers (SCTW), which was adopted in 1978 and amended in 1995.

It is up to ratifying nations to enforce these standards. The United States is a party to these conventions, and the U.S. Coast Guard enforces them when it boards and inspects ships and crews arriving at U.S. ports and the very few ships engaged in international trade that sail under the U.S. flag.

Like the United States, most of the other major maritime trading nations lack the ability to enforce these regulations as a "flag state" because much of the world's merchant fleet is registered under so-called "flags of convenience." While most ship owners and operators are headquartered in developed countries, they often register their ships in Panama, Liberia, the Bahamas, the Marshall Islands, Malta, and Cyprus, among other "open registries," because these nations offer more attractive tax and employment regulatory regimes. Because of this development, most maritime trading nations enforce shipping regulations under a "port state control" regime—that is, they require compliance with these regulations as a condition of calling at their ports.

The fragmented nature of ship ownership and operation can be a further hurdle to regulatory enforcement. It is common for cargo ships to be owned by one company, operated by a second company (which markets the ship's space), and managed by a third (which may supply the crew and other services a ship requires to sail), each of which could be headquartered in different countries.

Arctic Guidelines

While SOLAS and other IMO conventions include provisions regarding the operation of ships in ice-infested waters, they are not specific to the polar regions. To supplement existing requirements, in December 2002, the IMO approved guidelines for ships operating in Arctic ice-covered waters.[63] These were only recommendations for ships operating in the Arctic, not requirements. They apply to passenger and cargo ships of 500 gross tons or more engaged in international voyages. They do not apply to fishing vessels, military vessels, pleasure yachts, and smaller cargo ships. The guidelines are intended to improve safety and prevent pollution in the Arctic, and they include provisions on ship construction, ship equipment related to navigation, and crew training and ship operation. The guidelines recommend that ships carry fully enclosed lifeboats or carry tarpaulins to cover their lifeboats. They recommend that each crew include at least one ice navigator with documented evidence of having completed an ice navigation training program. The IMO is in the process of drafting mandatory requirements for ships operating in the Arctic.[64]

Nations can enforce additional requirements on ships arriving at their ports or sailing through their coastal waters. For instance, U.S. Coast Guard regulations largely follow IMO conventions but mandate additional requirements in some areas. U.S. coastal states can require ships calling at their ports to take additional safety and pollution prevention safeguards. Canada and Russia have additional pollution regulations for Arctic waters exceeding MARPOL. The U.S. Coast Guard is seeking agreement with Russia to establish a vessel traffic separation scheme for the Bering Strait between Alaska and Russia, which now experiences over 300 transits per year.[65]

Oil, Gas, and Mineral Exploration[66]

Changes to the Arctic brought about by warming temperatures resulting from climate change will likely allow more exploration for oil and gas offshore where the extent of summer sea ice is shrinking. Similarly, shrinking glaciers onshore could expose land containing economic deposits of gold, iron ore, or other minerals that were previously covered by glacial ice. Warming that causes permafrost to melt could also pose challenges to onshore exploration activities because ground structures, such as pipelines and other infrastructure that depend on footings sunk into the permafrost for support, could become unstable. Despite the warming temperatures, however, exploration and development in the Arctic would still be subject to harsh conditions, especially in winter, which makes it costly and challenging to develop infrastructure necessary to produce, store, and transport oil, gas, and minerals from newly discovered deposits.

Shrinking sea ice cover in the Arctic has also intensified interest in mapping the continental margins of countries with lands in the Arctic. Delineating the extent of the continental margins beyond the 200 nautical mile Exclusive Economic Zone (EEZ) under the terms of Article 76 of the U.N. Convention on the Law of the Sea (UNCLOS) could open up substantial amounts of submerged lands for development. Mapping projects are underway, by individual countries and in cooperative studies, that would be used to support national claims to submerged lands which may contain large amounts of oil, natural gas, methane hydrates, or minerals.

Expiration of the annual congressional moratoria on September 30, 2008, coupled with other developments in offshore leasing activity, impacts federal policies on Arctic offshore development in a number of ways.[67] The expiration of leasing restrictions allows Outer Continental Shelf (OCS) areas where leasing has not been considered for many years to be considered for potential federal leasing activity. Other developments in offshore leasing activity include a presidential order to lift the executive restrictions on certain OCS areas to allow offshore drilling,[68] the emergence of new offshore operations (OCS renewable energy leasing in Arctic areas),[69] and the general emergence of new technologies related to OCS research and development. Diminished leasing restrictions impact OCS activity in a domestic and an international context and generally contribute to the larger debate over OCS drilling in the Arctic.

Oil and Gas

A primary driver for the increased interest in exploring for oil and gas offshore in the Arctic is the shrinking Arctic ice cap, or conversely, the growing amount of ice-free ocean in the summertime. Reduced sea ice in the summer means that ships towing seismic arrays[70] can explore previously inaccessible regions of the Arctic Ocean, Chukchi Sea, Beaufort Sea and other offshore regions for longer periods of time without risk of colliding with floating sea ice. Less sea ice over longer periods compared to previous decades also means that the seasonal window for offshore drilling in the Arctic remains open longer in the summer and increases the chances for making a discovery.

In addition to the improved access to larger portions of the Arctic afforded by shrinking sea ice, recent interest in Arctic oil and gas was fueled by a 2008 U.S. Geological Survey (USGS) appraisal of undiscovered oil and gas north of the Arctic Circle.[71] The USGS asserts that "The extensive Arctic continental shelves may constitute the geographically largest unexplored prospective area for petroleum remaining on Earth."[72] In the report, the USGS estimates that 90 billion barrels of oil, nearly 1,700 trillion cubic feet of natural gas, and 44 billion barrels of natural gas liquids may remain to be discovered in the Arctic. Nearly all (84%) of the oil and gas is expected to occur offshore. The USGS estimate for total undiscovered oil and gas in the Arctic exceeds the total discovered amount of Arctic oil and oil-equivalent natural gas (240 billion barrels), which constitutes almost 10% of the world's known conventional petroleum resources.[73] An important caveat to the USGS assessment, however, is that it excludes economic considerations. The report only includes resources that would be technically recoverable with current technology; they may not necessarily be economically recoverable under prevailing oil and gas prices.

Despite the warming trend in the Arctic, a discovery of new oil and gas deposits far from existing storage, pipelines, and shipping facilities can not be developed until infrastructure is built to extract and transport the petroleum.

For example, a Devon Energy spokesman noted in April 2008 that a 240 million barrel discovery in the Beaufort Sea won't be developed for at least 10 years because the company does not have the infrastructure to transport the oil.[74] Other areas with oil and gas potential, such as offshore east Greenland, may take even longer to develop because they are still covered by substantial summer sea ice that would hinder exploration and development unless more ice melts.[75]

Extent of the Continental Margin

Increased interest in developing offshore resources in the Arctic has sparked efforts by nations bordering the Arctic Ocean to map the extent of their continental margins beyond the 200-mile EEZ limit. Under UNCLOS, nations can submit a claim to submerged lands if they demonstrate that their continental margin extends beyond the 200-mile limit according to the definition in Article 76. Under Article 76, the extent of the continental margin beyond the 200-mile limit depends on the position of the foot of the continental slope, the thickness of sediments, and the depth of water. Also, the continental margin could include geologic features that extend from the continent out to sea, which may include undersea ridges continuing for hundreds of miles offshore.

As mentioned earlier—see "U.S. Activities As a Non-Party to UNCLOS"—the United States, as a non-party to UNCLOS, cannot participate as a member of the Commission on the Limits of the Continental Shelf; it cannot submit a claim under Article 76. Over the years, however, it has submitted observations on submissions made by other states, requesting that those observations be made available online and to the commission. In addition, since 2001, the United States has gathered and analyzed data to determine the outer limits of its extended continental shelf.

Arctic border countries have begun the complex investigations needed to support claims for an extended continental shelf in the Arctic. Claims have already been submitted by several countries, including the Russian Federation, which submitted its UNCLOS claim to a portion of the Arctic continental shelf in 2001.[76] Russia made claims to the Lomonosov Ridge, an undersea feature spanning the Arctic from Russia to Canada, as an extension of its continental margin. The submission demonstrated Russia's bid to extend activities in Arctic regions. The United States has started to gather and analyze data through an initiative called the Extended Continental Shelf Project.[77] In this effort, the United States is working closely with Canada to prepare and present Canada's submission to the Commission on the Limits of the Continental Shelf.

Canada and the United States share overlapping regions of the seabed as part of the extended continental margin of both nations. Both countries have conducted research singly and jointly to map the extended continental shelf.[78] In August 2008, Canada's prime minister announced that Canada would spend $100 million (Canadian) over five years to map its total Arctic mineral and energy resources and bolster Canada's sovereignty over its northern resources.[79] On January 12, 2009, the Bush Administration issued a presidential directive addressing the extended continental shelf and boundary issues in the Arctic, among other issues, including national security and maritime transportation.[80] The directive acknowledges an unresolved boundary between the United States and Canada in the Beaufort Sea, and notes that Russia and the United States abide by a 1990 maritime boundary treaty pending its entry into force once it is ratified by the Russian Federation. According to the USGS, however, most of the potential oil and gas resources estimated for the Arctic are likely to occur within already agreed-upon territorial boundaries.[81] (For further discussion, see "Territorial Claims and Sovereignty Issues.")

Minerals

A warming Arctic means new opportunities and challenges for mineral exploration and development. Receding glaciers expose previously ice-covered land that could host economic mineral deposits that were previously undetectable and un-mineable below the ice. Longer summers would also extend exploration seasons even for areas not currently ice-covered but which are only accessible for ground surveys during the warmer months. In some parts of the Arctic, such as Baffin Island, Canada, less sea ice allows ships to transport heavy equipment to remote locations, and to convey ore from mines to the market further south. Some railway and mining operators are considering developing railroads and other infrastructure to transport ore year-round.[82] As with onshore oil and gas development, however, mining infrastructure that depends on footings sunk into permafrost could become unstable if the permafrost melts in response to warmer temperatures. Also, as with oil and gas development, mineral deposits that may be technically recoverable with current technology may not be economically profitable.

One important part of the current infrastructure in the Arctic that supports oil, gas, and mineral development is the construction and use of ice roads—built and used during the winter—but which are not passable during the warmer months. Warmer temperatures are shortening the ice road transport seasons and creating transportation challenges. For example, the opening date for tundra roads in northern Alaska has shifted from early November prior to 1991 to January in recent years.[83]

Environment and Economics

The warming Arctic has focused attention on threats to its wildlife and ecosystems, and it is expected that increased oil, gas, and mineral exploration and development activities may also invite increased scrutiny of possible harm to the fragile Arctic ecosystems. Federal offshore programs are often the subject of this type of scrutiny, as demonstrated by litigation largely focusing on environmental impacts. For example, plaintiffs in cases challenging government approval of OCS development activity in the Chukchi Sea generally are local communities and national environmental groups. The defendant in litigation over federal leasing in the Alaska program areas is the Secretary of the Interior. Typically litigation over federal offshore programs in Alaska takes place in two venues—the U.S. Court of Appeals for the Ninth Circuit, and/or the U.S. District Court of the District of Alaska. The schedule for litigation can be an issue because the schedule in one venue has some bearing on rulings related to litigation in the other venue. A new schedule for litigation regarding OCS development in the Chukchi Sea is currently under consideration. The outcome of court cases will likely impact future development activities.

The outcome of judicial action on environmental regulations is not yet known, but it is widely accepted that complying with current environmental laws and regulations in the Arctic adds to the comparatively high cost of doing business in the far north. In addition, the price and price stability of petroleum and mineral commodities will underlie industry decisions about whether to make large investments in multi-year exploration and development projects in the Arctic.

Low or volatile prices could curtail any sustained development efforts north of the Arctic Circle, even if continued warming opens new territory to petroleum and mineral exploration. In addition to uncertainty regarding the price of oil, gas, and minerals, and the environmental considerations that accompany exploration and development, there is the uncertainty for the United States regarding its claims to submerged lands beyond the 200-mile EEZ.

Even if the commodity prices and environmental costs were favorable toward exploration and development, uncertainty over U.S. claims to the extended continental shelf—because the

United States is a non-party to UNCLOS—may influence private sector decisions to invest in Arctic oil, gas, and mineral resource development.

Oil Pollution and Pollution Response[84]

Oil Pollution Implications of Arctic Change

Climate change impacts in the Arctic, particularly the decline of sea ice and retreating glaciers, have stimulated human activities in the region, many of which have the potential to create oil pollution. A primary concern is the threat of a large oil spill in the area. Although a major oil spill has not occurred in the Arctic region,[85] recent economic activity, such as oil and gas exploration and tourism (cruise ships), increases the risk of oil pollution (and other kinds of pollution) in the Arctic.

Significant spills in high northern latitudes (e.g., the 1989 *Exxon Valdez* spill in Alaska and spills in the North Sea) suggest that the "potential impacts of an Arctic spill are likely to be severe for Arctic species and ecosystems."[86]

Risk of Oil Pollution in the Arctic

A primary factor determining the risk of oil pollution in the Arctic is the level and type of human activity being conducted in the region. Although climate changes in the Arctic are expected to increase access to natural resources and shipping lanes, the region will continue to present logistical challenges that may hinder human activity in the region. For example (as discussed in another section of this report),[87] the unpredictable ice conditions may discourage trans-Arctic shipping. If trans-Arctic shipping were to occur on a frequent basis, it would represent a considerable portion of the overall risk of oil pollution in the region. In recent decades, many of the world's largest oil spills have been from oil tankers, which can carry millions of gallons of oil.[88]

Although the level of trans-Arctic shipping is uncertain, many expect oil exploration and extraction activities to intensify in the region.[89] Oil well blowouts from offshore oil extraction operations have been a source of major oil spills, eclipsing the largest tanker spills. The largest unintentional oil spill in recent history was from the 2010 *Deepwater Horizon* incident in the Gulf of Mexico.[90] During that incident, the uncontrolled well released (over an 84-day period) approximately 200 million gallons of crude oil into the Gulf.[91] The second-largest unintentional oil spill in recent history—the *IXTOC I*, estimated at 140 million gallons—was due to an oil well blowout in Mexican Gulf Coast waters in 1979.[92]

Until the 2010 *Deepwater Horizon* incident, the spill record for offshore platforms in U.S. federal waters had shown improvement from prior years.[93] A 2003 National Research Council (NRC) study of oil and gas activities on Alaska's North Slope stated "the conclusion of these analyses is that blowouts that result in large spills are unlikely."[94] Similar conclusions were made in federal agency documents regarding deepwater drilling in the Gulf of Mexico before the 2010 *Deepwater Horizon* event.[95] Some would likely contend that the underlying analyses behind these conclusions should be adjusted to account for the 2010 Gulf oil spill. However, others may argue that the proposed activities in U.S. Arctic waters present less risk of an oil well blowout than was encountered by the *Deepwater Horizon* drill rig, because the proposed U.S. Arctic operations would be in shallower waters (150 feet) than the deepwater well (approximately 5,000 feet) that was involved in the 2010 Gulf oil spill. In addition, Shell Oil has stated that the pressures in the Chukchi Sea (the location of Shell's recent interest) would be two to three times less than they were in well involved in the 2010 Gulf oil spill.[96] Regardless of these differences, even under the most stringent control systems, some oil spills and other accidents are likely to occur from equipment failure or human error.

Potential Impacts

No oil spill is entirely benign. Even a relatively minor spill, depending on the timing and location, can cause significant harm to individual organisms and entire populations. Regarding aquatic spills, marine mammals, birds, bottom-dwelling and intertidal species, and organisms in early developmental stages—eggs or larvae—are especially vulnerable. However, the effects of oil spills can vary greatly. Oil spills can cause impacts over a range of time scales, from only a few days to several years, or even decades in some cases.

Conditions in the Arctic may have implications for toxicological effects that are not yet understood. For example, oil spills on permafrost may persist in an ecosystem for relatively long periods of time, potentially harming plant life through their root systems. Moreover, little is known about the effects of oil spills on species that are unique to the Arctic, particularly, species' abilities to thrive in a cold environment and the effect temperature has on toxicity.[97]

The effects of oil spills in high latitude, cold ocean environments may last longer and cause greater damage than expected. Some recent studies have found that oil spills in lower latitudes have persisted for longer than initially expected, thus raising the concern that the persistence of oil in the Arctic may be understated. In terms of wildlife, population recovery may take longer in the Arctic because many of the species have longer life spans and reproduce at a slower rate.[98]

Response and Cleanup Challenges in the Arctic Region

Climate changes in the Arctic are expected to increase human activities in the region, many of which impose a risk of oil pollution, particularly from oil spills. Conditions in the Arctic region impose unique challenges for personnel charged with (1) oil spill response, the process of getting people and equipment to the incident, and (2) cleanup duties, either recovering the spilled oil or mitigating the contamination so that it poses less harm to the ecosystem. These challenges may play a role in the policy development for economic activities in the Arctic.

Spill Response Challenges

Response time is a critical factor for oil spill recovery. With each hour, spilled oil becomes more difficult to track, contain, and recover, particularly in icy conditions, where oil can migrate under or mix with surrounding ice.[99] Most response techniques call for quick action, which may pose logistical challenges in areas without prior staging equipment or trained response professionals. Many stakeholders are concerned about a "response gap" for oil spills in the Arctic region.[100] A response gap is a period of time in which oil spill response activities would be unsafe or infeasible. The response gap for the northern Arctic latitudes is likely to be extremely high compared to other regions.[101]

According to the former Commander of the 17th Coast Guard District (Alaska), "we are not prepared for a major oil spill [over 100,000 gallons] in the Arctic environment. The Coast

Guard has no offshore response capability in Northern or Western Alaska."[102] The transportation infrastructure along Alaska's northern coast poses challenges for oil spill responders. The Coast Guard has no designated air stations north of Kodiak, AK, which is almost 1,000 miles from the northernmost point of land along the Alaskan coast in Point Barrow, AK.[103] Although some of the communities have airstrips capable of landing cargo planes, no roads connect these communities.[104] Vessel infrastructure is also limited. The nearest major port is in the Aleutian Islands, approximately 1,300 miles from Point Barrow. A 2010 Government Accountability Office (GAO) report identified further logistical obstacles that would hinder an oil spill response in the region, including "inadequate" ocean and weather information for the Arctic and technological problems with communications.[105]

The history of oil spills and response in the Aleutian Islands raises concerns for potential spills in the Arctic region:

> The past 20 years of data on response to spills in the Aleutians has also shown that almost no oil has been recovered during events where attempts have been made by the responsible parties or government agencies, and that in many cases, weather and other conditions have prevented any response at all.[106]

Oil Spill Cleanup Challenges

The behavior of oil spills in cold and icy waters is not well understood.[107] Cleaning up oil spills in ice-covered waters will be more difficult than in other areas, primarily because effective strategies have yet to be developed. Natural oil seeps, which are a major source of oil in the arctic environment, may offer opportunities for studying the behavior of oil.[108]

The Arctic conditions present several hurdles to oil cleanup efforts. In colder water temperatures, there are fewer organisms to break down the oil through microbial degradation. Oil evaporates at a slower rate in colder water temperatures. Although slower evaporation may allow for more oil to be recovered, evaporation removes the lighter, more toxic hydrocarbons that are present in crude oil.[109] The longer the oil remains in an ecosystem, the more opportunity there is for exposure. Oil spills may get trapped in ice, evaporating only when the ice thaws. In some cases, oil could remain in the ice for years.

Icy conditions enhance emulsification—the process of forming different states of water in oil, often described as "mousse." Emulsification creates oil cleanup challenges by increasing (1) the volume of the oil/water mixture and (2) the mixture's viscosity (resistance to flow). The latter change creates particular problems for conventional removal and pumping cleanup methods.[110] Moreover, two of the major non-mechanical recovery methods—in-situ burning and dispersant application—may be limited (or "precluded") by the Arctic conditions and lack of logistical support: aircraft, vessels, and other infrastructure.[111]

Existing Policy Framework

Considering both the recent increase in human activity in the region (and expectation of further interest) and the response and recovery challenges that an oil spill would impose in Arctic waters, many would assert that the region warrants particular attention in terms of governance. However, the existing framework for international governance of maritime operations in the Arctic region lacks legally binding requirements. While the Safety of Life at Sea Convention (SOLAS) and other International Maritime Organization (IMO) conventions include provisions regarding ships in icy waters, the provisions are not specific to the polar regions. Although the IMO has "Guidelines for Ships Operating in Arctic," a 2009 NOAA report described the non-binding IMO provisions as "inconsistent with the hazards of Arctic navigation and the potential for environmental damage from such an incident."[112]

Fisheries[113]

Large commercial fisheries exist in the Arctic, including in the Barents and Norwegian Seas north of Europe, the Central North Atlantic off of Greenland and Iceland, and the Newfoundland and Labrador Seas off of northeastern Canada.[114]

In the 110th Congress, P.L. 110-243 stated that "the United States should initiate international discussions and take necessary steps with other Arctic nations to negotiate an agreement or agreements for managing migratory, transboundary, and straddling fish stocks in the Arctic Ocean and establishing a new international fisheries management organization or organizations for the region." In response to this, the United States participated in meetings with other Arctic nations at the March 2009 U.N. Food and Agriculture Organization Committee on Fisheries meetings in Rome. These international discussions are continuing. No additional legislation has been enacted related to U.S. fisheries in the Arctic.

On February 5, 2009, the North Pacific Fishery Management Council, acting under the authority of the Magnuson-Stevens Fishery Conservation and Management Act, unanimously approved the Council's draft Arctic Fisheries Management Plan and approved the preferred alternative, which would (1) close the Arctic to commercial fishing until information improves so that fishing can be conducted sustainably and with due concern to other ecosystem components; (2) determine the fishery management authorities in the Arctic and provide the Council with a vehicle for addressing future management issues; and (3) implement an ecosystem-based management policy that recognizes the resources of the U.S. Arctic and the potential for fishery development that might affect those resources, particularly in the face of a changing climate.

On November 3, 2009, the National Marine Fisheries Service (Department of Commerce) issued a final rule that implemented the North Pacific Council's Fishery Management Plan for Fish Resources of the Arctic Management Area, as well as Amendment 29 to the Fishery Management Plan for Bering Sea/Aleutian Islands King and Tanner Crabs. This action followed the North Pacific Council's recommendations establishing a basis for sustainable management of commercial fishing in the Arctic Management Area and moving the northern boundary of the crab fishery south of the Arctic Management Area to Bering Strait.[115] In October 2009, an international symposium focused on conservation of fish stocks and the potential for future fisheries management in the Arctic Ocean.[116] The National Marine Fisheries Service maintains an "Arctic Fisheries" web page summarizing activates related to fisheries.[117]

Protected Species[118]

Concern over development of the Arctic relates to how such development might affect threatened and endangered species. Under the Endangered Species Act (ESA, 16 U.S.C. §§1531-1543), the polar bear was listed as threatened on May 15, 2008. The failure by the

Fish and Wildlife Service (FWS) to make a 90-day finding on a 2008 petition to list Pacific walrus led to submission of 60- days' notice of a future citizen suit. However, eventually walruses were listed as candidate species under ESA;[119] this status means that federal agencies carrying out actions that may affect the species must confer with FWS though they are not necessarily obliged to modify their actions. Both polar bears and walruses are heavily dependent during their life cycles on thick sea ice, making them especially susceptible to the shrinking Arctic ice cap.

On December 30, 2008, the National Marine Fisheries Service (NMFS) determined that a listing of ribbon seal as threatened or endangered was not warranted.[120] On October 22, 2010, NMFS listed the southern distinct population segment (DPS) of spotted seals as threatened.[121] Listing of two other DPS (Okhotsk and Bering Sea) had earlier been determined to not be warranted.[122] On December 10, 2010, NMFS proposed that (1) four subspecies of ringed seal be listed as threatened,[123] and (2) that two DPS of one subspecies of bearded seal be listed as threatened.[124]

In either terrestrial or marine environments, the extreme pace of change makes a biological response many times more difficult. For species with adaptations for a specific optimum temperature for egg development, or production of young timed to match the availability of a favored prey species, or seed dispersal in predictable fire regimes, etc., evolutionary responses may well not keep pace with the rate of change.[125] While species of plants and animals farther south might migrate, drift, or be transplanted from warming habitats to more northerly sites that may continue to be suitable,[126] once a terrestrial species reaches the Arctic Ocean, it is very literally at the end of the line. No more northern or colder habitat is available.

The Marine Mammal Protection Act (MMPA; 16 U.S.C. §§1361 et seq.) protects whales, seals, walruses, and polar bears. The MMPA established a moratorium on the "taking" of marine mammals in U.S. waters and by U.S. nationals on the high seas, including the Arctic. The MMPA protects marine mammals from "clubbing, mutilation, poisoning, capture in nets, and other human actions that lead to extinction." Under the MMPA, the Secretary of Commerce, acting through National Marine Fisheries Service (NMFS), is responsible for the conservation and management of whales and seals. The Secretary of the Interior, acting through the Fish and Wildlife Service (FWS), is responsible for walruses and polar bears.[127] Despite the MMPA's general moratorium on taking, the MMPA allows U.S. citizens to apply for and obtain authorization for taking small numbers of mammals incidental to activities other than commercial fishing (e.g., offshore oil and gas exploration and development) if the taking would have only a negligible impact on any marine mammal species or stock, provided that monitoring requirements and other conditions are met.

Indigenous People Living in the Arctic[128]

People have been living in the Arctic for thousands of years, and indigenous peoples developed highly specialized cultures and economics based on the physical and biological conditions of the long-isolated region. However, with trade, the influx of additional populations especially since the 19th century, and ongoing physical changes in the Arctic, indigenous populations have already experienced substantial change in their lifestyles and economies. Over the past two decades, greater political organization across indigenous populations has increased their demands for international recognition and broader rights, as well as attention to the economic, health, and safety implications of climate change in the North.

Background

Seven of the eight Arctic nations have indigenous peoples,[129] whose predecessors were present in parts of the Arctic over 10,000 years ago, well before the arrival of peoples with European backgrounds.[130] Current Arctic indigenous peoples comprise dozens of diverse cultures and speak dozens of languages from eight or more non-Indo-European language families.[131]

Before the arrival of Europeans, Arctic indigenous peoples lived in economies that were chiefly dependent, in varying proportions, on hunting land and marine mammals, catching salt- and fresh-water fish, herding reindeer (in Eurasia), and gathering, for their food, clothing, and other products.[132] Indigenous peoples' interaction with and knowledge of Arctic wildlife and environments has developed over millennia and is the foundation of their cultures.[133]

The length of time that Arctic indigenous peoples were in contact with Europeans varied across the Arctic. As recorded by Europeans, contact began as early as the 9th century CE, if not before, in Fennoscandia[134] and northwestern Russia, chiefly for reasons of commerce (especially furs); it progressed mostly west-to-east across northern Asia, reaching northeastern Arctic Asia by the 17th century.[135] North American Arctic indigenous peoples' contact with Europeans started in Labrador in the 16th century and in Alaska in the 18th century, and was not completed until the early 20th century.[136] Greenland's indigenous peoples first saw European-origin peoples in the late 10th century, but those Europeans died out during the 15th or 16th century and Europeans did not return permanently until the 18th century.[137]

Contact led to significant changes in Arctic indigenous economies, political structures, foods, cultures, and populations, starting especially in the 20th century. For example, life expectancy among Alaska Natives has increased from 47 years in 1950 to over 69 years in 2000 (though it still lags behind that of U.S. residents overall, at 77 years).[138]

Also, at present, most Arctic indigenous peoples have become minorities in their countries' Arctic areas, except in Greenland and Canada. (One source estimates that, around 2003, about 10% of an estimated 3.7 million people in the Arctic were indigenous.[139]) While many Arctic indigenous communities remain heavily dependent on hunting, fishing, and herding and are more likely to depend on traditional foods than non-indigenous Arctic inhabitants,[140] there is much variation. Most Arctic indigenous people may no longer consume traditional foods as their chief sources of energy and nutrition.[141] Major economic change is also relatively recent but ongoing.[142] Many Arctic indigenous communities have developed a mixture of traditional economic activities and wage employment.[143] The economics of subsistence and globalization will be key factors in the effects of climate change on Arctic indigenous peoples, and on their reactions to Arctic climate change.

Arctic indigenous peoples' current political structures vary, as do their relationships with their national governments. Some indigenous groups govern their own unique land areas within the national structure, as in the United States and Canada; others have special representative bodies, such as the Saami parliaments in Norway, Finland, and Sweden;[144] a few areas have general governments with indigenous majorities, such as Greenland (a member country of Denmark), Nunavut territory in Canada, and the North Slope and Northwest Arctic boroughs in Alaska.[145] Control of land, through claims and ownership, also varies among Arctic indigenous peoples, as do rights to fishing, hunting, and resources.[146] Arctic indigenous peoples' political relationships to their national and local governments, and their ownership or claims regarding land, are also significant factors in the responses to Arctic climate change by the indigenous peoples and by Arctic nations' governments.

Effects of Climate Change

Arctic climate change is expected to affect the economies, population, subsistence, health, infrastructure, societies, and cultures of Arctic indigenous peoples. Changes in sea ice and sea level, permafrost, tundra, weather, and vegetation distributions, as well as increased commercial shipping, mineral extraction, and tourism, will affect the distribution of land and sea mammals, of freshwater and marine fish, and of forage for reindeer. These will in turn affect traditional subsistence activities and related indigenous lifestyles.[147] Arctic indigenous peoples' harvesting of animals is likely to become riskier and less predictable, which may increase food insecurity, change diets, and increase dependency on outside, non-traditional foods.[148] Food cellars in many locations have thawed during summers, threatening food safety. Related health risks of diabetes, obesity, and mental illness have been associated with these changes.[149]

Sea, shoreline ice, and permafrost changes have damaged infrastructure and increased coastal and inland erosion, especially in Alaska, where GAO found in 2003 that "coastal villages are becoming more susceptible to flooding and erosion caused in part by rising temperatures."[150] In response, Congress funded the U.S. Army Corps of Engineers to conduct a Baseline Erosion Assessment that identified and prioritized among the 178 communities identified at risk from erosion.[151] (Risks from flooding were not examined.) GAO concluded in 2009 that many Native villages must relocate, but even those facing imminent threats have been impeded by various barriers, including difficulties identifying appropriate new sites, piecemeal programs for state and federal assistance, and obstacles to eligibility for certain federal programs.[152] The Alaska Federation of Natives placed among its 2010 federal priorities a request to Congress to mitigate flooding and erosion in Alaska Native villages and to fund relocation of villages where necessary.[153] However, "the cost is extraordinary," acknowledges Senator Lisa Murkowski.[154]

Oil, gas, and mineral exploration and development are expected to increase, as are other economic activities, such as forestry and tourism, and these are expected to increase economic opportunities for all Arctic residents, including indigenous peoples.[155] Pressures to increase participation in the wage economy, however, may speed up changes in indigenous cultures. Increased economic opportunities may also lead to a rise in the non-indigenous population, which may further change the circumstances of indigenous cultures. Some representatives of Arctic indigenous people have related a "conflicting desire between combating climate change and embracing the potential for economic growth through foreign investment."[156]

Although important advances in public health have occurred in indigenous communities over past decades, some health problems may increase with continued Arctic climate change. Economic development may exacerbate Arctic pollution problems, including higher exposure to mercury, air pollution, and food contamination. The influx and redistribution of contaminants in the air, oceans, and land may change in ways that are now poorly understood.[157] Warmer temperatures and longer warm seasons may increase insect- and wildlife-borne diseases.[158] Climate change may lead to damage to water and sanitation systems, reducing protection against waterborne diseases.[159] Changes in Arctic indigenous cultures may increase mental stress and behavioral problems.[160]

The response to climate change by Arctic indigenous peoples has included international activities by Arctic indigenous organizations and advocacy before their national governments. As one report noted, "the rise of solidarity among indigenous peoples organizations in the region is surely a development to be reckoned with by all those interested in policy issues in the Arctic."[161] Six national or international indigenous organizations are permanent participants of the Arctic Council, the regional intergovernmental forum.[162] Due in part to advocacy by Arctic indigenous people, the United Nations General Assembly adopted in 2007 the Declaration on the Rights of Indigenous Peoples.[163] In April 2009, the Inuit Circumpolar Council (an organization of Inuit in the Arctic regions of Alaska, Canada, Greenland, and Russia) hosted in Alaska the worldwide "Indigenous Peoples Global Summit on Climate Change."[164] The conference report, forwarded to the Copenhagen Conference of the Parties of the U.N. Framework Convention on Climate Change (December 2009), noted "accelerating" climate change caused by "unsustainable development" and, among several recommendations, called for a greater indigenous role in national and international decisions on climate change, including a greater role for indigenous knowledge in climate change research, monitoring, and mitigation.[165]

Polar Icebreaking[166]

The Coast Guard's polar icebreakers perform a variety of missions supporting U.S. interests in the Arctic (and Antarctic), including the following:

- conducting and supporting U.S. scientific research in the Arctic;[167]
- defending U.S. sovereignty in the Arctic by helping to maintain a presence in the region;
- defending other U.S. interests in the Arctic, including economic interests relating to the U.S. exclusive economic zone (EEZ) north of Alaska;
- monitoring sea traffic in the Arctic, including ships bound for the United States; and
- conducting other typical Coast Guard missions (such as search and rescue, law enforcement, and protection of marine resources) in Arctic waters, including U.S. territorial waters north of Alaska.

The Coast Guard's two existing heavy polar icebreakers—*Polar Star* and *Polar Sea*—have exceeded their originally intended 30-year service lives. *Polar Star* was placed in caretaker status on July 1, 2006. Congress in FY2009 and FY2010 provided funding to repair it and return it to service for an additional 7 to 10 years of service; the repair work was completed and the ship was reactivated on December 14, 2012.

On June 25, 2010, the Coast Guard announced that *Polar Sea* had suffered an unexpected engine casualty; the ship was unavailable for operation after that. The Coast Guard placed *Polar Sea* in commissioned, inactive status on October 14, 2011. Section 222 of the Coast Guard and Maritime Transportation Act of 2012 (H.R. 2838/P.L. 112-213 of December 20,

2012) prohibits the Coast Guard from removing any part of *Polar Sea* and from transferring, relinquishing ownership of, dismantling, or recycling the ship until it submits a business case analysis of the options for and costs of reactivating the ship and extending its service life to at least September 30, 2022, so as to maintain U.S. polar icebreaking capabilities and fulfill the Coast Guard's high latitude mission needs, as identified in the Coast Guard's July 2010 High Latitude Study.

The Coast Guard's third polar icebreaker—*Healy*—entered service in 2000. Compared to *Polar Star* and *Polar Sea*, Healy has less icebreaking capability (it is considered a medium polar icebreaker), but more capability for supporting scientific research. The ship is used primarily for supporting scientific research in the Arctic.

The Coast Guard's proposed FY2013 budget includes $8 million in acquisition funding to initiate survey and design activities for a new polar icebreaker. The Coast Guard's Five Year Capital Investment Plan includes an additional $852 million in FY2014-FY2017 for acquiring the ship. The Coast Guard anticipates awarding a construction contract for the ship "within the next five years" and taking delivery on the ship "within a decade." The project to design and build a polar icebreaker is a new acquisition project initiated in the FY2013 budget.

With the reactivation of *Polar Star*, the operational U.S. polar icebreaking fleet consists of one heavy polar icebreaker (*Polar Star*) and one medium polar icebreaker (*Healy*). The new polar icebreaker for which initial acquisition funding is requested in the FY2013 budget would replace *Polar Star* at about the time Polar Star's 7- to 10-year reactivation period ends.

In July 2011, the Coast Guard provided to Congress a study on the Coast Guard's missions and capabilities for operations in high-latitude (i.e., polar) areas. The study, commonly known as the High Latitude Study and dated July 2010 on its cover, concluded the following:

- **The Coast Guard requires three heavy and three medium icebreakers to fulfill its statutory missions.** These icebreakers are necessary to (1) satisfy Arctic winter and transition season demands and (2) provide sufficient capacity to also execute summer missions. Single-crewed icebreakers have sufficient capacity for all current and expected statutory missions. Multiple crewing provides no advantage because the number of icebreakers required is driven by winter and shoulder season requirements. Future use of multiple or augmented crews could provide additional capacity needed to absorb mission growth.
- **The Coast Guard requires six heavy and four medium icebreakers to fulfill its statutory missions and maintain the continuous presence requirements of the Naval Operations Concept.** Consistent with current practice, these icebreakers are single-crewed and homeported in Seattle Washington.
- **Applying crewing and home porting alternatives reduces the overall requirement to four heavy and two medium icebreakers.** This assessment of non-material solutions shows that the reduced number of icebreakers can be achieved by having all vessels operate with multiple crews and two of the heavy icebreakers homeporting in the Southern Hemisphere.

Leasing was also considered as a nonmaterial solution. While there is no dispute that the Coast Guard's polar icebreaker fleet is in need of recapitalization, the decision to acquire this capability through purchase of new vessels, reconstruction of existing ships,

> or commercial lease of suitable vessels must be resolved to provide the best value to the taxpayer. The multi-mission nature of the Coast Guard may provide opportunities to conduct some subset of its missions with non government-owned vessels. However, serious consideration must be given to the fact that the inherently governmental missions of the Coast Guard must be performed using government-owned and operated vessels. An interpretation of the national policy is needed to determine the resource level that best supports the nation's interests....
>
> The existing icebreaker capacity, two inoperative heavy icebreakers and an operational medium icebreaker, does not represent a viable capability to the federal government. The time needed to augment this capability is on the order of 10 years. At that point, around 2020, the heavy icebreaking capability bridging strategy expires.[168]

At a July 27, 2011, hearing on U.S. economic interests in the Arctic before the Oceans, Atmosphere, Fisheries, and Coast Guard subcommittee of the Senate Commerce, Science, and Transportation Committee, the following exchange occurred:

> SENATOR OLYMPIA J. SNOWE: On the high latitude study, do you agree with—and those—I would like to also hear from you, Admiral Titley, as well, on these requirements in terms of Coast Guard vessels as I understand it, they want to have—I guess, it was a three medium ice breakers. Am in correct in saying that? Three medium ice breakers.
>
> ADMIRAL ROBERT PAPP, COMMANDANT OF THE COAST GUARD: I agree with the mission analysis and as you look at the requirements for the things that we might do up there, if it is in the nation's interest, it identifies a minimum requirement for three heavy ice breakers and three medium ice breakers and then if you want a persistent presence up there, it would require—and also doing things such as breaking out (inaudible) and other responsibilities, then it would take up to a maximum six heavy and four medium.
>
> SNOWE: Right. Do you agree with that?
>
> PAPP: If we were to be charged with carrying out those full responsibilities, yes, ma'am. Those are the numbers that you would need to do it.
>
> SNOWE: Admiral Titley, how would you respond to the high latitude study and has the Navy conducted its own assessment of its capability?
>
> REAR ADMIRAL DAVID TITLEY, OCEANORGRAPHER AND NAVIGATOR OF THE NAVY: Ma'am, we are in the process right now of conducting what we call a capabilities based assessment that will be out in the summer of this year.
>
> We are getting ready to finish that—the Coast Guard has been a key component of the Navy's task force on climate change, literally since day one when the Chief of Naval Operations set this up, that morning, we had the Coast Guard invited as a member of our executive steering committee.
>
> So we have been working very closely with the Coast Guard, with the Department of Homeland Security, and I think Admiral Papp—said it best as far as the specific comments on the high latitude study but we have been working very closely with the Coast Guard.[169]

Potential issues for Congress regarding Coast Guard polar icebreaker modernization include the following:

- the numbers and capabilities of polar icebreakers the Coast Guard will need in the future;
- the disposition of *Polar Sea*;

- whether the new polar icebreaker initiated in the FY2013 budget should be funded with incremental funding (as proposed in the Coast Guard's Five Year Capital Investment Plan) or full funding in a single year, as required under the executive branch's full funding policy;
- whether new polar icebreakers should be funded entirely in the Coast Guard budget, or partly or entirely in some other part of the federal budget, such as the Department of Defense (DOD) budget, the National Science Foundation (NSF) budget, or both;
- whether to provide future icebreaking capability through construction of new ships or service life extensions of existing polar icebreakers; and whether future polar icebreakers should be acquired through a traditional acquisition or a leasing arrangement.

Search and Rescue[170]

General

The possibility of increased sea and air traffic through Arctic waters has increased concerns regarding Arctic-area search and rescue capabilities. Given the location of current U.S. Coast Guard operating bases, it could take Coast Guard aircraft several hours, and Coast Guard cutters days or even weeks, to reach a ship or a downed aircraft in distress in Arctic waters. In addition, the harsh climate complicates search and rescue operations in the region. Particular concern has been expressed about cruise ships that may experience problems and need assistance; there have already been incidents of this kind in recent years in waters off Antarctica. A May 7, 2011, press report stated:

> U.S. and Canadian military commanders say they are examining their rescue capabilities in the Arctic as a shrinking ice cap brought about by climate change opens up rich oil and gas reserves and draws more commercial traffic to the top of the globe....
>
> [U.S. Admiral James Winnefeld, commander of the U.S. Northern Command and the North American Aerospace Defense Command, or NORAD] and NORAD'S deputy commander, Canadian Forces Lt. Gen. Marcel Duval, said more ships in the Arctic Ocean - and more jetliners crossing Arctic skies - could mean more emergencies.
>
> "More ships, more chances of accidents," Duval said in a separate interview in his office at NORAD headquarters at Peterson.
>
> Last summer, the Canadian Coast Guard rescued 197 people from the cruise ship Clipper Adventurer, which ran aground inside the Arctic Circle....
>
> Sea traffic is still light, with only about 25 ships a year currently crossing the maritime Arctic boundary between Alaska and the Yukon. But that number is increasing by 10 to 15 percent a year, according to NORAD statistics.
>
> Civilian air traffic over the Arctic is booming. U.S. and Canadian aviation agencies report more than 9,600 civilian flights across the North Pole in 2010, up nearly 21 percent from 2008.[171]

One option for mitigating the risks associated with cruise ship operations in the Arctic would be to require the vessels to sail in pairs, so that if one ship experiences a problem, the other one could quickly come to its aid.

Increasing U.S. Coast Guard search and rescue capabilities for the Arctic could require one or more of the following: enhancing or creating new Coast Guard operating bases in the region; procuring additional Arctic-capable aircraft, cutters, and rescue boats for the Coast Guard; and adding systems to improve Arctic maritime communications, navigation, and domain awareness.[172] It may also entail enhanced forms of cooperation with navies and coast guards of other Arctic countries.

Coast Guard Statements

A January 2011 press report summarized remarks made by the Commandant of the Coast Guard, Admiral Robert Papp, concerning Coast Guard search and rescue capabilities in the Arctic. The article quoted Papp as saying that there is a need for a range of Coast Guard operational capability for the Arctic, and that "In the meantime, he said the service will lean on partnerships with other Arctic nations. However, he warned that the Coast Guard will likely not be able to respond to any crises in the Arctic circle in a timely fashion. He recalled that the Canadian Coast Guard came under fire when it took six days to rescue a cruise ship and oil tanker that both ran aground in its northern waters." The article stated:

> "We wouldn't be able to make it in six days," he said. "It'd probably take us six weeks to get adequate resources up for a similar thing in our waters, so we have to start focusing on this."[173]

A March 2011 press report summarizing remarks made by Admiral Papp during and after a hearing before the Coast Guard and Maritime Transportation subcommittee of the House Transportation and Infrastructure Committee stated:

> Because retreating ice continues to make the passageways in the Arctic Ocean more navigable, the Coast Guard needs to have air stations in the region to conduct helicopter rescue missions, Papp said. He has not chosen specific cities for the air station sites, but he said the Alaskan cities of Barrow and Kotzebue come to mind.
>
> When lawmakers focus on Arctic resources, the United States' shortage of icebreakers usually draws the bulk of attention, but Papp said the Coast Guard is also in dire need of small boats to conduct rescue missions.[174]

At a June 23, 2011, hearing on the Coast Guard's proposed FY2012 budget before the Oceans, Atmosphere, Fisheries, and Coast Guard subcommittee of the Senate Commerce, Science, and Transportation Committee, the following exchange occurred:

> ADMIRAL ROBERT J. PAPP, JR., COMMANDANT OF THE COAST GUARD: I visited Barrow, Kotzebue, and Nome. And, actually, it was a revisit because I had served up there as an ensign (ph) 35 years ago. And so it was good to get back up there and see the changes.
>
> But what has not changed is the infrastructure up there. And I think that we have to have a robust discussion on the infrastructure needed to support what is no doubt going to be an increase in human activity up there off the – the North Coast of Alaska.
>
> Icebreakers I think are important, but they cloud the discussion of the other needs that we have up there. And I think we've focused on icebreakers over the last few years, even though they are important...
>
> SENATOR MARK BEGICH: Right.
>
> PAPP: ... and needed. But, right now, if we were to have to mount a response like we did in the Gulf of Mexico—I sent 3,000 people down for Deepwater Horizon. You know how many hotel rooms are available in Barrow.

BEGICH: That's right.

PAPP: And we—we have no place to put people up there. We have no hangars for aircraft. We have no piers, no Coast Guard boats. So my immediate, pressing concern is as human activity occurs, as you have that ship that goes through that—first of all, we'll assure safety standards because no matter where ships operate in the world, we are involved in their safety standards.

But if an accident happens, how do we respond? And, right now, we've got zero capability to respond in the Arctic right now. And we've got to do better than that.

That—when people ask me what keeps me awake at night—an oil spill, a collision, a ship sinking in the Arctic keeps me awake at night because we have nothing to respond or, if we respond, it's going to take us weeks to get there.[175]

State Department Statement

The State Department states that:

> the United States and the seven other Arctic Council (AC) Member States (Canada, Denmark, Finland, Iceland, Norway, Russian Federation, Sweden) adopted a Ministerial Declaration in April 2009 in Tromso, Norway, which established a Task Force with the mandate of developing a Search and Rescue (SAR) agreement for the Arctic. The United States hosted the first meeting of the Arctic SAR Task Force during December 9-11, 2009, in Washington, D.C. The next round of negotiations is scheduled [sic] for February 25-26, 2010, in Moscow, Russian Federation. The Arctic SAR Task Force has been asked to finalize the Arctic SAR agreement such that it can be presented for adoption by the AC at its Ministerial meeting in Spring 2011.
>
> The United States is at the forefront of efforts to promote safety in the Arctic. The Department of State's Office of Ocean and Polar Affairs is coordinating federal interagency interest to negotiate an instrument for the saving of lives at sea and the rescue of survivors after aircraft accidents in the Arctic. The U.S. Coast Guard is a prominent agency participating in this effort. With so few resources available for SAR in the Arctic, developing a regional agreement to set baseline standards for greater international cooperation and coordination will be of great value.[176]

Arctic Council Agreement on Arctic Search and Rescue (May 2011)

On May 12, 2011, representatives from the member states of the Arctic Council, meeting in Nuuk, Greenland, signed an agreement on cooperation on aeronautical and maritime search and rescue in the Arctic. Secretary of State Hillary Rodham Clinton signed for the United States. Key features of the agreement include the following:

- Article 2 states: "The objective of this Agreement is to strengthen aeronautical and maritime search and rescue cooperation and coordination in the Arctic."
- Article 3 and the associated Annex to the agreement essentially divide the Arctic into search and rescue areas within which each party has primary responsibility for conducting search and rescue operations, stating that "the delimitation of search and rescue regions is not related to and shall not prejudice the delimitation of any boundary between States or their sovereignty, sovereign rights or jurisdiction," and that "each Party shall promote the establishment, operation and maintenance of an adequate and effective search and rescue capability within its area."

- Article 4 and the associated Appendix I to the agreement identify the competent authority for each party. For the United States, the competent authority is the Coast Guard.
- Article 5 and the associated Appendix II to the agreement identify the agencies responsible for aeronautical and maritime search and rescue for each party. For the United States, those agencies are the Coast Guard and the Department of Defense.
- Article 6 and the associated Appendix III to the agreement identify the aeronautical and/or maritime rescue coordination centers (RCCs) for each party. For the United States, the RCCs are Joint Rescue Coordination Center Juneau (JRCC Juneau) and Aviation Rescue Coordination Center Elmendorf (ARCC Elmendorf).
- Article 7 states that "the Parties shall conduct aeronautical and maritime search and rescue operations pursuant to this Agreement consistent" with certain general guidelines.[177]
- Article 8 states that "a Party requesting permission to enter the territory of a Party or Parties for search and rescue purposes, including for refueling, shall send its request to a search and rescue agency and/or RCC of the relevant Party or Parties," and that "the Party receiving such a request shall immediately confirm such receipt. The receiving Party, through its RCCs, shall advise as soon as possible as to whether entry into its territory has been permitted and the conditions, if any, under which the mission may be undertaken."
- Article 9 states that "the Parties shall enhance cooperation among themselves in matters relevant to this Agreement," that "the Parties shall exchange information that may serve to improve the effectiveness of search and rescue operations," and that "the Parties shall promote mutual search and rescue cooperation by giving due consideration to collaborative efforts."
- Article 10 states that "the Parties shall meet on a regular basis in order to consider and resolve issues regarding practical cooperation."
- Article 11 states that "after a major joint search and rescue operation, the search and rescue agencies of the Parties may conduct a joint review of the operation led by the Party that coordinated the operation."
- Article 12 states that "unless otherwise agreed, each Party shall bear its own costs deriving from its implementation of this Agreement," and that "implementation of this Agreement shall be subject to the availability of relevant resources."
- Article 18 states that "any Party to this Agreement may, where appropriate, seek cooperation with States not party to this Agreement that may be able to contribute to the conduct of search and rescue operations, consistent with existing international agreements."
- Article 19 states that "any Party may at any time withdraw from this Agreement by sending written notification thereof to the depositary[178] through diplomatic channels at least six months in advance."[179]

Figure 4 shows an illustrative map of the national areas of search and rescue responsibility based on the geographic coordinates listed in the Annex to the agreement.

The State Department states that the agreement:

is the first legally-binding instrument negotiated under the auspices of the Arctic Council. It coordinates life-saving international maritime and aeronautical SAR coverage and response among the Arctic States across an area of about 13 million square miles in the Arctic.

As Arctic sea ice coverage decreases, ship-borne activities are increasing significantly in the Arctic. Flight traffic is also on the rise as new polar aviation routes cross the Arctic air space in several directions. As human presence and activities in the Arctic expand, the potential for accidents increases as well. Limited rescue resources, challenging weather conditions, and the remoteness of the area render SAR operations difficult in the Arctic, making coordination among the Arctic nations imperative. The SAR Agreement will improve search and rescue response in the Arctic by committing all Parties to coordinate appropriate assistance to those in distress and to cooperate with each other in undertaking SAR operations. For each Party, the Agreement defines an area of the Arctic in which it will have lead responsibility in organizing responses to SAR incidents, both large and small. Parties to the Agreement commit to provide SAR assistance regardless of the nationality or status of persons who may need it.

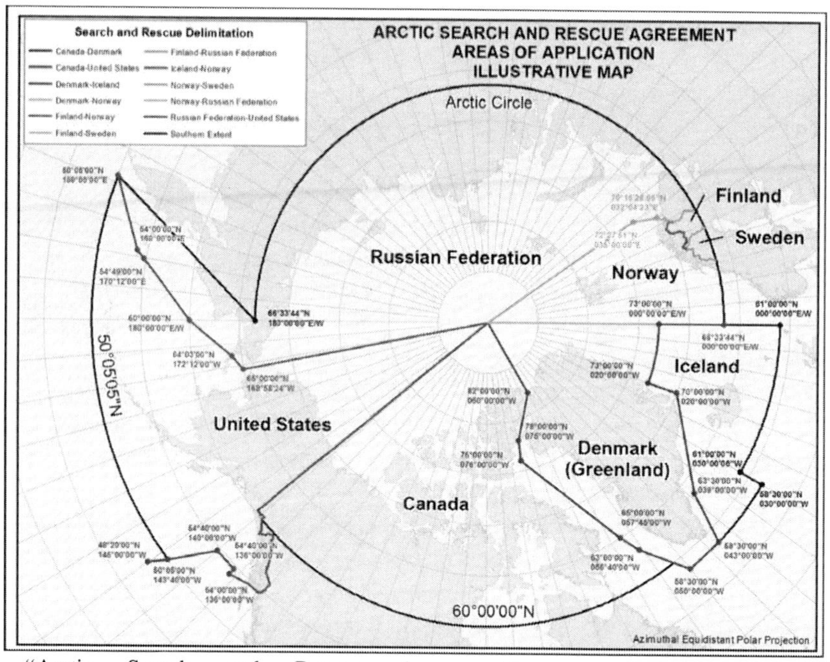

Source: "Arctic Search and Rescue Agreement," accessed July 7, 2011 at http://www.arcticportal.org/features/ features-of-2011/arctic-search-and-rescue-agreement.

Figure 4. Illustrative Map of Arctic SAR Areas in Arctic SAR Agreement; (Based on geographic coordinates listed in the agreement).

The Arctic Council launched this initiative at its 2009 Ministerial Meeting in Tromso, Norway, establishing a Task Force, co-chaired by the United States and the Russian Federation. The Task Force proceeded in a highly collaborative spirit, meeting five times (in Washington, Moscow, Oslo, Helsinki and Reykjavik).

The signature of the SAR Agreement in Nuuk is a positive step toward building partnerships in the Arctic. In particular, it reflects the commitment of the Arctic Council

States to enhance their cooperation and offer responsible assistance to those involved in accidents in one of the harshest environments on Earth.

This Agreement illustrates one of the most successful negotiations to date to address emerging issues in the Arctic. Arctic Council participants approached SAR negotiations with collaboration and dedication to a positive outcome. The United States congratulates its colleagues in this effort and looks forward to further collaboration on the vital issues facing the rich but fragile Arctic region.[180]

Geopolitical Environment[181]

Many observers have noted that the loss of Arctic ice[182] is already leading to stepped-up human activity in the high north, particularly in the form of increasing commercial traffic and development. This trend has brought forth a range of issues on the geopolitical front, from environmental protection to search-and-rescue capabilities to the delineation of national boundaries—which will determine access to natural resources. These concerns are being addressed cooperatively in both bilateral and multilateral fashion, especially under the aegis of the Arctic Council and the U.N. Convention on the Law of the Sea (UNCLOS). Nonetheless, some observers continue to raise questions concerning security in the high north, and have advanced sometimes conflicting views regarding the potential roles of military forces in the region. Of the other Arctic coastal nations, the United States enjoys strong political and commercial ties with Canada, Norway, and Denmark; all four countries are members of NATO. Although the United States views Russia as an important partner in developing policies to cope with changing conditions in the Arctic, relations with Moscow have been somewhat problematic. Several non-Arctic nations, including India and China, have also evinced interest in gaining permanent observer status in the Arctic Council. In addition, the European Union, which sought but was denied full observer status in 2009, is nonetheless developing policy toward the Arctic.[183]

Multilateral Cooperation

As noted elsewhere in this report (see "Territorial Claims and Sovereignty Issues"), in late May 2008, ministerial representatives of the five Arctic littoral states attended a meeting convened by Denmark in Ilulissat, Greenland (a semi-autonomous territory of Denmark). Danish Foreign Minister Stig Møller implied that the meeting was intended to develop interim measures for Arctic governance: "We must continue to fulfill our obligations in the Arctic area until the U.N. decides who will have the right to the sea and the resources in the region. We must agree on the rules and what to do if climate changes make more shipping possible."[184] Attendees discussed a variety of issues, including the environment, transportation, resources, and security. The Danish Foreign Ministry stated that the resulting Ilulissat Declaration

> sent a clear political signal to the local inhabitants and the rest of the world that we will act responsibly when addressing the development in the Arctic Ocean. We have committed ourselves politically to solve any disagreements through negotiation. Thus, hopefully, we have eradicated all the myths about a 'race for the North Pole.' The legal framework is in place and the five States have now declared that they will abide by it.[185]

The Arctic has increasingly become a subject of discussions in bilateral meetings among leaders of the nations in the region. The main international forum for cooperation in the high north, however, is the eight-nation Arctic Council, of which the United States is an active member. The chief topics addressed by the six working groups of the Council, which was formed in 1996, are sustainable development, environmental protection, and the social well-being of the indigenous communities. The United States vetoed security as an issue of consideration for the Council. The Council holds ministerial-level meetings biennially, while the working groups meet more frequently. Sweden currently holds the two-year revolving Council chairmanship; Canada will take over in 2013, followed by the United States.[186]

On May 11-12, 2011, Secretary of State Hillary Clinton attended the most recent Arctic Council ministerial summit, held in Nuuk, Greenland; she was accompanied by U.S. Interior Secretary Ken Salazar. They were the first U.S. Cabinet members to attend an Arctic Council meeting, and observers noted that their visits served to raise the profile of Arctic issues. The meeting was led by Sweden, which currently chairs the Council. Noting the increased commercial activity in the region, Secretary Clinton stated, "We need to pursue these opportunities in a smart, sustainable way that preserves the Arctic environment and ecosystem."[187] Among other issues, attendees focused on efforts to reduce emissions that cause "black carbon" to settle on the Arctic region, accelerating ice melt. In addition, the Council discussed launching a longer-term study on methods to cope with possible future oil spills.

The major "deliverable" of the summit, however, was the signing of an Agreement on Cooperation on Aeronautical and Maritime Search and Rescue (SAR) in the Arctic. The accord was hailed as the "first legally-binding instrument negotiated under the auspices of the Arctic Council." The SAR initiative, developed mainly by the United States and Russia, had been introduced during the April 2009 summit. The Council also approved the establishment of a permanent secretariat, to be based in Tromsø, Norway.[188]

In early June 2012, Secretary Clinton traveled once more to the Arctic, visiting the new secretariat in Tromsø as part of an eight-day trip to Scandinavia. While there, she emphasized that the United States "want[s] the Arctic Council to remain the premier institution that deals with Arctic questions."[189]

Analysts note that Russia is keen to capitalize on natural resource development and shipping in the Arctic. As noted elsewhere in this report, Russia and Norway in late 2010 resolved a 40-year dispute over national borders in the Barents Sea; the accord will permit exploration for undersea oil, believed to be in rich supply there. In addition, scientists estimate that the sea route along the Siberian coast (referred to as the Northeast Passage or the Northern Sea Route, or NSR) will be ice-free and navigable well before the Northwest Passage through the Canadian archipelago; 34 vessels sailed the NSR in 2011. Russia sees significant economic opportunities in offering icebreaker escorts, refueling posts, and supplies to the commercial ships that will use the waterway.[190] But Russian leaders also understand that doing so will require international cooperation and goodwill. In August 2012, Russia is scheduled to join the United States and Norway in the military exercise *Northern Eagle 2012*.[191]

Although its borders lie some distance from the Arctic, China has also displayed a growing interest in the region, based mainly upon the potential opportunities for shorter sea routes and the eventual development of energy-related natural resources. China's economy is strongly dependent upon exports; some analysts have estimated that as much as one-half of

China's GDP is reliant upon exports and shipping. China also is reliant upon ocean transportation for its large petroleum imports. The opening of Arctic sea corridors north of Russia and/or Canada would drastically reduce both sailing times and transportation costs. Beijing is keenly interested in having free access to these future waterways.

Many Chinese believe that the Arctic should be considered as part of the global commons. According to David Curtis Wright of the Naval War College, "The mantra that the Arctic and its natural resource wealth belong to no one country but constitute the common heritage of all humankind is virtually de rigueur in recent Chinese public commentary on Arctic affairs."[192] During a May 2012 workshop on Sino-Nordic cooperation hosted in Beijing, Chinese participants referred to their country as a "near-Arctic state" and a "stakeholder." Some analysts believe that China will likely remain officially circumspect on this question, as its "foreign policy rests on a profound respect for territorial integrity."[193] Nonetheless, some Chinese analysts reportedly are encouraging the government to challenge Canada's claim of sovereignty over the Northwest Passage. China has also been active in conducting research on the Arctic; it acquired a large icebreaker in 1993 and has constructed a state-of-the-art polar capable research vessel, the *Snow Dragon*. In April 2012, Chinese Premier Wen Jiabao visited Sweden and Iceland, and two months later President Hu Jintao went to Denmark; the two leaders were reportedly discussing large-scale investments in the region. In addition, China (like several other nations) has established a research station in the Svalbard archipelago and has beefed up the size of its embassy staff in Iceland.

Security Issues

Throughout the Cold War, the Arctic region was a zone of strategic interest, where the United States, the Soviet Union, and allied states conducted air and naval maneuvers and tested ballistic missiles. With the collapse of the Warsaw Pact and the Soviet Union, however, the importance of the high north diminished in the 1990s. Although the establishment of sovereignty through the demarcation of boundaries in the region is being conducted peacefully under the auspices of the U.N. Convention on the Law of the Sea (UNCLOS), the Arctic is once again being viewed by some as a potential emerging security issue. In a December 2011 *Washington Post* op-ed, Heather Conley, a senior fellow at the Center for Strategic and International Studies, noted several recent developments:

> In April [2011], President Obama signed a new command plan that gives NORAD and the U.S. Northern Command greater responsibility in protecting the North Pole and U.S. Arctic territory.... In 2009, Norway moved its operational command to its northern territories above the Arctic Circle. Russia has plans to establish a brigade that is specially equipped and prepared for military warfare in Arctic conditions. Denmark has made it a strategic priority to form an Arctic Command. Canada is set to revitalize its Arctic fleet, including spending $33 billion to build 28 vessels over the next 30 years.[194]

Canadian academic Rob Huebert has pointed out that in August 2010 the United States, Canada, and Denmark conducted in the Canadian Arctic their annual joint naval exercises involving several advanced and powerful warships. Huebert observed that "while defence officials are quick to point out they see no military threat to the region, it's still interesting to see these three Arctic friends coming together to improve their naval combat capability in the Far North."[195] In varying degrees, the Arctic coastal states have indicated a willingness to establish and maintain a military presence in the high north.[196] Although some have argued that terrorism and hijacking may constitute security concerns in the region, others maintain that such threats are chimerical, given the challenges of distance and geography, and the difficulty of navigating in a polar environment. *The Economist* has asserted that "the risks of Arctic conflict have been exaggerated. Most of the Arctic is clearly assigned to individual countries. According to a Danish estimate, 95% of Arctic mineral resources are within agreed national boundaries."[197]

The Arctic has also become a region of interest for NATO. However, as one writer has noted, "[t]here is currently no consensus within the alliance that NATO has any role to play in the Arctic, as Canada strongly opposes any NATO involvement on sovereignty grounds and other NATO members are concerned with negative Russian reaction."[198] Speaking in Reykjavík in January 2009, former NATO Secretary General Jaap de Hoop Scheffer urged that member states not allow the Arctic to become a divisive issue. He also recommended that the alliance and Russia cooperate through building upon their shared experience in search-and-rescue operations. Former Danish Prime Minister Anders Fogh Rasmussen, who became secretary general of the alliance in August 2009, has also addressed security in the high north. Citing the "potentially huge security implications" of Arctic climate change, Rasmussen in October 2009 stated that "I think it is within the natural scope of work for NATO to be the forum for consultation and discussion on [selected Arctic] issues."[199] In March 2009, however, Russia's NATO ambassador stated that Moscow would not cooperate with the alliance on Arctic matters. And in September 2010, President Medvedev reportedly observed that "the Arctic can do fine without NATO," and that his government "views [possible NATO] activity with quite serious tension, because it is after all a zone of peaceful cooperation, economic cooperation, and of course the military factor always—at a minimum—creates additional questions." On a visit to Moscow in November 2010, Rasmussen assured the Russians that NATO does not intend to establish a presence in the Arctic.[200]

Since 2006, several member and partner states have participated in *Cold Response*, a wide-ranging annual crisis response exercise hosted by Norway. During the most recent joint maneuvers in March 2012, 14 nations fielded air, land, and naval assets and more than 16,000 troops. Although the exercises are multilateral, they are not conducted under the auspices of NATO.[201]

It has been noted that Russia "has at least half of the Arctic in terms of area, coastline, population and probably mineral wealth."[202] The Russian government has stated that, although it deplores the notion of an arms race in the region and does not foresee a conflict there, it intends to protect its Arctic interests.[203] However, Russia has at times appeared to be sending out mixed messages.

For example, at the conclusion of a meeting in September 2010, Russian Foreign Minister Sergei Lavrov and Canadian Foreign Minister Lawrence Cannon stated that "any militarization [of the Arctic] is out of the question." And in June 2011, then-Prime Minister (and currently President) Vladimir Putin stated, "Russia will definitely expand its presence in the Arctic. We are open for dialogue with our foreign partners and with all neighbors in the Arctic region. But we will naturally defend our own geopolitical interests firmly and consistently." The following month, Putin announced plans to build a large shipping port on the Yamal peninsula, and the government stated that it would be sending two brigades to the north to protect its interests.

Similarly, some Chinese leaders have voiced concern over perceived emerging security issues in the Arctic. In early March 2010, a Chinese admiral stated that "the current scramble for the sovereignty of the Arctic among some nations has encroached on many other countries' interests," and he added that China had to "make short and long term ocean strategic development plans to exploit the Arctic because it will become a future mission for the navy." Some analysts, however, believe that China's general approach toward the Arctic will remain decidedly low-key: "To date, China has adopted a wait-and-see approach to Arctic developments, wary that active overtures would cause alarm in other countries due to China's size and status as a rising global power."

China is believed to be keen on resolving through diplomacy the national interests of both littoral and non-Arctic states in the high north. Toward that end, it has sought permanent observer status on the Arctic Council. However, its candidacy may be imperiled by a dispute with Norway, which in 2010 awarded the Nobel Peace Prize to Chinese dissident Liu Xiaobo. Norwegian political scientist Gunhild Hoogense Gjørv has noted that "[e]veryone is interested in the moves that China is making."[204]

As noted, the Arctic Council does not address regional security issues. To fill this apparent void, a recent report by the Center for Strategic and International Studies has proposed the creation of a separate organization, the Arctic Coast Guard Forum (ACGF), consisting initially of the eight Arctic Council states, but possibly expanding eventually to include other countries willing to contribute assets.

The ACGF, which could potentially be headquartered at the U.S. Air Force base in Thule, Greenland, would "focus first on information sharing yet should also seek to develop methods of cooperation in support of the Arctic Council's search-and-rescue agreement and future international oil spill response agreement."[205]

U.S. Military Forces and Operations[206]

During the Cold War, the Arctic was an arena of military competition between the United States and the Soviet Union, with both countries, for example, operating nuclear-powered submarines, long-range bombers, and tactical aircraft in the region.[207] The end of the Cold War and the collapse of most elements of the Russian military establishment following the dissolution of the Soviet Union in December 1991 greatly reduced this competition and led to a reduced emphasis on the Arctic in U.S. military planning.

The diminishment of Arctic sea ice is now leading U.S. military forces to pay renewed attention to the Arctic. This is particularly true in the case of the Navy and Coast Guard, for whom diminishment of Arctic sea ice is opening up potential new operating areas for their surface ships.

Navy and Coast Guard activities relating to the Arctic are taking place as other countries, such as Canada, Russia, and Norway, are examining the potential implications for their military forces of diminished Arctic sea ice, and taking or contemplating steps to increase their own navy and coast guard presence and operations in the region.[208] Defense officials in the United States and other countries view issues such as sovereignty, freedom of navigation, and energy exploration as creating a potential in the Arctic for military cooperation, competition, or conflict, depending on how these issues are handled.[209]

DOD in General

2010 Quadrennial Defense Review

The Department of Defense's (DOD's) report on the 2010 Quadrennial Defense Review (QDR), submitted to Congress in February 2010, states:

> The effect of changing climate on the Department's operating environment is evident in the maritime commons of the Arctic. The opening of the Arctic waters in the decades ahead[,] which will permit seasonal commerce and transit[,] presents a unique opportunity to work collaboratively in multilateral forums to promote a balanced approach to improving human and environmental security in the region. In that effort, DoD must work with the Coast Guard and the Department of Homeland Security to address gaps in Arctic communications, domain awareness, search and rescue, and environmental observation and forecasting capabilities to support both current and future planning and operations. To support cooperative engagement in the Arctic, DoD strongly supports accession to the United Nations Convention on the Law of the Sea.[210]

The report also states:

> The Department of Defense and its interagency partners must be able to more comprehensively monitor the air, land, maritime, space, and cyber domains for potential direct threats to the United States. Such monitoring provides the U.S. homeland with an extended, layered in depth defense. This effort includes enhanced coordination with Canada for the defense of North America as well as assisting Mexico and Caribbean partners in developing air and maritime domain awareness capacities. Special attention is required to develop domain awareness tools for the Arctic approaches as well. In coordination with domestic and international partners, DoD will explore technologies that have the potential to detect, track, and identify threats in these spheres to ensure that capabilities can be deployed to counter them in a timely fashion.[211]

The report further states:

> Central to the security of the United States is a strong transatlantic partnership, which is underpinned by the bilateral relationships between the United States and the governments of Europe. We will continue to work with this community of like-minded nations, whether by engaging with allies still shaping their democracies after decades of living in the shadow of the Soviet Union, building on the benefits of French reintegration into NATO's military structure, or addressing new security issues such as those arising in the Arctic region.[212]

The report states: "We will seek out opportunities to work with Moscow on emerging issues, such as the future of the Arctic" and that DOD "will also enhance defense relationships and continue to work with Canada in the context of regional security, increased interaction in the Arctic, and combat operations in Afghanistan."[213]

April 2011 Change to Unified Command Plan

In April 2011, it was reported that

> Changes made to the U.S. military's Unified Command Plan shift geographic boundaries and stress the growing importance of the Arctic, officials said.
>
> President Barack Obama signed the document yesterday [April 6].

The biggest change to the plan assigns U.S. Northern Command responsibility for the Arctic. U.S. European Command and U.S. Pacific Command shared responsibility with U.S. Northern Command for the region under the last change published in December 2008. It also places responsibility for Alaska under Northern Command. The previous plan had Northern Command and U.S. Pacific Command sharing responsibility for the state and adjacent waters.

"Northcom was given advocacy responsibility for Arctic capabilities primarily due to having the only U.S. Arctic territory within its area of operations," a Pentagon spokesman said.

Northern Command also already works closely with Canada and "has a habitual relationship with the Department of Homeland Security and the U.S. Coast Guard," the spokesman continued. "These relationships are key to human and environmental safety and security."...

The Unified Command Plan is the responsibility of the chairman of the Joint Chiefs of Staff, and is reviewed every two years. The Joint Staff coordinates input from the combatant commanders, the service chiefs and Defense Department leadership. The chairman, Navy Adm. Mike Mullen, submitted his recommendations through Defense Secretary Robert M. Gates to Obama.[214]

May 2011 Report to Congress

A May 2011 DOD report to Congress on Arctic operations and the Northwest Passage that was done at congressional direction[215] stated:

> The Arctic is warming on average twice as fast as the rest of the planet, resulting in increased human activity in the region. Although some perceive that competition for resources and boundary disputes may result in conflict in the Arctic, the opening of the Arctic also presents opportunities to work collaboratively in multilateral forums to promote a balanced approach to improving human and environmental security in the region.
>
> Strategic guidance on the Arctic is articulated in National Security Presidential Directive (NSPD) 66 / Homeland Security Presidential Directive (HSPD) 25, *Arctic Region Policy*. Additional guidance is found in *the 2010 National Security Strategy (NSS)* and the *2010 Quadrennial Defense Review (QDR)*. The overarching strategic national security objective is *a stable and secure region where U.S. national interests are safeguarded and the U.S. homeland is protected.* This objective is consistent with a regional policy that reflects the relatively low level of threat in a region bounded by nation states that have not only publicly committed to working within a common framework of international law and diplomatic engagement, but also demonstrated ability and commitment to doing so over the last fifty years.
>
> DoD will take responsible steps to anticipate and prepare for the Arctic operations of the near-(2010-2020), mid- (2020-2030), and far-term (beyond 2030). Capabilities will need to be reevaluated as conditions change, and gaps must be addressed in order to be prepared to operate in a more accessible Arctic. Key challenges include: shortfalls in ice and weather reporting and forecasting; limitations in command, control, communications, computers, intelligence, surveillance, and reconnaissance (C4ISR) due to lack of assets and harsh environmental conditions; limited inventory of ice-capable vessels; and limited shore-based infrastructure. The key will be to address needs in step with the rate at which activity in the Arctic increases, and balance potential investments in these capabilities with other national priorities. The United States has a vital Arctic neighbor and partner in Canada, with its shared values and interests in the region. DoD will work with the Canadian Department of National Defence (DND) to ensure common Arctic interests are addressed in a complementary manner.

Although having multiple Combatant Commanders (CCDRs) in the Arctic Ocean makes coordination more challenging, having too few would leave out key stakeholders, diminish longstanding relationships, and potentially alienate important partners. There are now two CCDRs with Arctic responsibilities: Commander, U.S. European Command (USEUCOM), and Commander, U.S. Northern Command (USNORTHCOM), each responsible for a portion of the Arctic Ocean aligned with adjacent land boundaries, an arrangement judged best suited to achieve continuity of effort with key regional partners.

Existing DoD posture in the region is adequate to meet near- to mid-term U.S. defense needs. DoD does not currently anticipate a need for the construction of a deep-draft port in Alaska between now and 2020. Given the long lead times for construction of major infrastructure in the region, DoD will periodically re-evaluate this assessment as the Combatant Commanders update their regional plans on a regular basis.

The United States needs assured Arctic access to support national interests in the Arctic. This access can be provided by a variety of proven capabilities, including submarines and aircraft, but only U.S.-flagged ice-capable ships provide visible U.S. sovereign maritime presence throughout the Arctic region. This need could potentially be met by either icebreakers or ice-strengthened surface vessels, none of which are in the U.S. Navy current surface combatant inventory, but which do exist in U.S. Coast Guard's inventory in limited numbers.

Finally, significant uncertainty remains about the rate and extent of climate change in the Arctic and the pace at which human activity will increase. The challenge is to balance the risk of being late-to-need with the opportunity cost of making premature Arctic investments. Not only does early investment take resources from other pressing needs, but the capabilities would be later in their lifecycle when finally employed. Given the many competing demands on DoD's resources in the current fiscal environment, the Department believes that further evaluation of the future operating environment is required before entertaining significant investments in infrastructure or capabilities.[216]

January 2012 GAO Report Reviewing May 2011 DOD Report

A January 2012, congressionally directed[217] GAO report reviewing the May 2011 DOD report above stated the following:

While DOD has undertaken some efforts to assess the capabilities needed to meet national security objectives in the Arctic, it is unclear whether DOD will be in a position to provide needed capabilities in a timely and efficient manner because it lacks a risk-based investment strategy for addressing near-term needs and a collaborative forum with the Coast Guard for addressing long-term capability needs. DOD's [May 2011] Arctic Report acknowledges that it has some near-term gaps in key capabilities needed to communicate, navigate, and maintain awareness of activity in the region. However, DOD has not yet evaluated, selected, or implemented alternatives for prioritizing and addressing near-term Arctic capability needs. In addition, DOD and the Coast Guard have established a working group to identify potential collaborative efforts to enhance U.S. Arctic capabilities. This working group is focused on identifying potential near-term investments but not longer-term needs, and it is currently expected to be dissolved in January 2012. Uncertainty involving the rate of Arctic climate change necessitates careful planning to ensure efficient use of resources in developing Arctic needs such as basing infrastructure and icebreakers, which require long lead times to develop and are expensive to build and maintain. Without taking steps to meet near- and long-term Arctic capability needs, DOD risks making premature Arctic investments, being late in obtaining needed capabilities, or missing opportunities to minimize costs by collaborating on investments with the Coast Guard....

DOD has several efforts under way to assess the capabilities needed to support U.S. strategic objectives in the Arctic. However, it has not yet developed a comprehensive approach to addressing Arctic capabilities that would include steps such as developing a risk-based investment strategy and timeline to address near-term needs and establishing a collaborative forum with the Coast Guard to identify long-term Arctic investments....

According to DOD's Arctic Report, capabilities will need to be reassessed as conditions change, and gaps will need to be addressed to be prepared to operate in a more accessible Arctic. Other stakeholders have also assessed Arctic capability gaps. Examples of these efforts include the following:

- U.S. Northern Command initiated a commander's estimate for the Arctic in December 2010 that, according to officials, will establish the commander's intent and missions in the Arctic and identify capability shortfalls. In addition, Northern Command identified two Arctic-specific capability gaps (communications and maritime domain awareness) in its fiscal years 2013 through 2017 integrated priority list, which defines the combatant command's highest-priority capability gaps for the near-term, including shortfalls that may adversely affect missions.
- U.S. European Command completed an Arctic Strategic Assessment in April 2011 that, among other things, identified Arctic capability gaps in the areas of environmental protection, maritime domain awareness, cooperative development of environmental awareness technology, sharing of environmental data, and lessons learned on infrastructure development. In addition, it recommended that the command conduct a more detailed mission analysis for potential Arctic missions, complete a detailed capability estimate for Arctic operations, and work in conjunction with Northern Command and the Departments of the Navy and Air Force to conduct a comprehensive capabilities-based assessment for the Arctic.
- DOD and DHS established the Capabilities Assessment Working Group (working group) in May 2011 to identify shared Arctic capability gaps as well as opportunities and approaches to overcome them, to include making recommendations for near-term investments. The working group was directed by its Terms of Reference to focus on four primary capability areas when identifying potential collaborative efforts to enhance Arctic capabilities, including near-term investments. Those capability areas include maritime domain awareness, communications, infrastructure, and presence. The working group was also directed to identify overlaps and redundancies in established and emerging DOD and DHS Arctic requirements. As the advocate for Arctic capabilities, Northern Command was assigned lead responsibility for DOD in the working group, while the Coast Guard was assigned lead responsibility for DHS. The establishment of the working group—which, among other things, is to identify opportunities for bi-departmental action to close Arctic capability gaps and issue recommendations for near-term investments—helps to ensure that collaboration between the Coast Guard and DOD is taking place to identify near-term capabilities needed to support current planning and operations. Although the working group is developing a paper with its recommendations, officials indicated that additional assessments would be required to address those recommendations.
- U.S. Navy completed its first Arctic capabilities-based assessment in September 2011 and is developing a second capabilities-based assessment focused on observing,

mapping, and environmental prediction capabilities in the Arctic, which officials expect to be completed in the spring of 2012. The Navy's first Arctic capabilities-based assessment identified three critical capability gaps as the highest priorities, including the capabilities to provide environmental information; maneuver safely on the sea surface; and conduct training, exercise, and education. This assessment recommended several near-term actions to address these gaps....

Even though DOD has made preliminary efforts to identify Arctic capability gaps and assess strategic objectives, constraints, and risks in the Arctic, DOD has not yet evaluated, selected, or implemented alternatives for prioritizing and addressing near-term Arctic capability needs....

Given that the opening in the Arctic presents a wide range of challenges for DOD, a risk-based investment strategy and timeline can help DOD develop the capabilities needed to meet national security interests in the region. Without a risk-based investment strategy and timeline for prioritizing and addressing near-term Arctic capability gaps and challenges, which is periodically updated to reflect evolving needs, DOD could be slow to develop needed capabilities, potentially facing operational risk and higher costs if the need arises to execute plans rapidly. Conversely, DOD could move too early, making premature Arctic investments that take resources from other, more pressing needs or producing capabilities that could be outdated before they are used.

While DOD and DHS have established the working group to identify shared near-term Arctic capability gaps, this collaborative forum is not intended to address long-term Arctic capability gaps or identify -opportunities for joint investments over the longer-term. DOD acknowledged the importance of collaboration with the Coast Guard over the long-term in its 2010 Quadrennial Defense Review, which states that the department must work with the Coast Guard and DHS to develop Arctic capabilities to support both current and future planning and operations. According to DOD and Coast Guard officials, although the working group is primarily focused on near-term investments, it has discussed some mid- to longterm capability needs. However, DOD and Coast Guard officials stated that after the completion of the working group's paper, expected in January 2012, the working group will have completed the tasks detailed in the Terms of Reference and will be dissolved. Consequently, no forum will exist to further address any mid- to long-term capability needs....

After the working group completes its tasks in January 2012, there will be no DOD and Coast Guard organization focused specifically on reducing overlap and redundancies or collaborating to address Arctic capability gaps in support of future planning and operations, as is directed by the 2010 Quadrennial Defense Review.

While Northern Command officials stated they have plans for periodic reassessment of longterm capability needs, such as icebreakers or basing infrastructure including a deep-water port, it is not clear how those plans consider collaboration with the Coast Guard. For example, officials stated the biennial review of Northern Command's Theater Campaign Plan and Strategic Infrastructure Master Plan will consider long-term capability and infrastructure needs. They added that the commander's Arctic Estimate is reviewed annually and also considers long-term priorities, such as identifying a need for icebreakers. However, the officials stated that the Arctic

Estimate does not identify how DOD would acquire those icebreakers or how it would coordinate with the Coast Guard—the operator of the nation's icebreakers—to reconstruct existing or build new icebreakers....

Without specific plans for a collaborative forum between DOD and the Coast Guard to address long-term Arctic capability gaps and to identify opportunities for joint investments over the longer-term, DOD may miss opportunities to leverage resources with the Coast Guard to enhance future Arctic capabilities.

At this time, significant DOD investments in Arctic capabilities may not be needed, but that does not preclude taking steps to anticipate and prepare for Arctic operations in the future. Addressing near-term gaps is essential for DOD to have the key enabling capabilities it needs to communicate, navigate, and maintain awareness of activity in the region. An investment strategy that identifies and prioritizes near-term Arctic capability needs and identifies a timeline to address them would be useful for decision makers in planning and budgeting. Without taking deliberate steps to analyze risks in the Arctic and prioritize related resource and operational requirements, DOD could later find itself faced with urgent needs, resulting in higher costs that could have been avoided.

In addition, unless DOD and DHS continue to collaborate to identify opportunities for interagency action to close Arctic capability gaps, DOD could miss out on opportunities to work with the Coast Guard to leverage resources for shared needs. DOD may choose to create a new collaborative forum or incorporate this collaboration into an existing forum or process. Given the different missions and associated timelines of DOD and the Coast Guard for developing Arctic capabilities, it is important that the two agencies work together to avoid fragmented efforts and reduce unaffordable overlap and redundancies while addressing Arctic capability gaps in support of future planning and operations.

To more effectively leverage federal investments in Arctic capabilities in a resource-constrained environment and ensure needed capabilities are developed in a timely way, we recommend that the Secretary of Defense, in consultation with the Secretary of the Department of Homeland Security, take the following two actions:

- develop a risk-based investment strategy that: 1) identifies and prioritizes near-term Arctic capability needs, 2) develops a timeline for addressing them, and 3) is updated as appropriate; and
- establish a collaborative forum with the Coast Guard to fully leverage federal investments and help avoid overlap and redundancies in addressing long-term Arctic capability needs.

In written comments on a draft of this report, DHS concurred with both of our recommendations. For its part, DOD partially concurred with both of our recommendations. It generally agreed that the department needed to take action to address the issues we raised but indicated it is already taking initial steps to address them.[218]

June 2012 Senate Armed Services Committee Report Language

The Senate Armed Services Committee, in its report (S.Rept. 112-173 of June 4, 2012) on S. 3254, the FY2013 National Defense Authorization Act, stated:

Arctic Region

The committee recognizes the continued importance of the Arctic region to our broader national strategy. Declining ice cover continues to open the Arctic region and a concerted, systematic, and immediate effort should be undertaken to adequately protect the United States' security, environmental, energy, economic, and natural resource interests in the Arctic.

The committee commends the Department of Defense (DOD) and, in particular, U.S. Northern Command and the Department of the Navy, and other federal agencies, such as the Department of Homeland Security (DHS), through the U.S. Coast Guard, the Department of Commerce, through the National Oceanic and Atmospheric Administration and the U.S. Arctic Research Commission, for their progress thus far in studying the region and defining the capabilities required to effectively operate in and protect the domain.

International and interagency collaboration is also needed to develop the necessary mapping and charting resources required for safe navigation and to promote security and economic interests. As sea ice recedes, timely weather forecasts and disaster warnings along with more baseline data will be required to conduct successful search and rescue missions. Search and rescue coordination, planning, and training for the Arctic should be thoroughly analyzed and developed to ensure forces can successfully operate in the domain.

The committee recognizes the importance of DOD's involvement in interagency and international efforts to protect national security interests in the region and, accordingly, urges DOD to continue to work in concert with DHS to establish a formal chartered working group to pursue increasing Arctic capabilities in the areas of communications, maritime domain awareness, infrastructure, and presence, as was recommended by U.S. Northern Command and the U.S. Coast Guard in a joint White Paper dated March 13, 2012. The committee also urges the agencies to develop an investment strategy for funding emerging requirements in balance with a resource constrained environment. (Pages 194-195)

Navy and Coast Guard

The Navy and Coast Guard are exploring the potential implications that increased surface ship and aircraft operations in the Arctic may have for required numbers of ships and aircraft, ship and aircraft characteristics, new or enlarged Arctic bases, and supporting systems, such as navigation and communication systems. The Navy and Coast Guard have sponsored or participated in studies and conferences to explore these implications, the Coast Guard has deployed boats and aircraft into the region to better understand the implications of operating such units there,[219] and Navy sailors have ridden on Canadian Navy ships deploying to Arctic waters for similar reasons.[220]

Points or themes that have emerged in studies, conferences, and deployments regarding the potential implications for the U.S. Navy and Coast Guard of diminished Arctic sea ice include but are not limited to the following:

- The diminishment of Arctic ice is creating potential new operating areas in the Arctic for Navy surface ships and Coast Guard cutters.
- U.S. national security interests in the Arctic include "such matters as missile defense and early warning; deployment of sea and air systems for strategic sealift, strategic deterrence, maritime presence, and maritime security operations; and ensuring freedom of navigation and overflight."[221]

- A mission of potential particular interest for expanded surface ship operations in the Arctic would be defending the U.S. (and European Union) claim that the Northern Sea Route running along Russia's north coast and the Northwest Passage running through Canada's northern archipelago constitute international straits which allow right of innocent passage.
- Search and rescue in the Arctic is a mission of increasing importance, particularly for the Coast Guard, and one that poses potentially significant operational challenges (see "Search and Rescue" above).
- More complete and detailed information on the Arctic is needed to more properly support expanded Navy and Coast Guard ship and aircraft operations in the Arctic.
- The Navy and the Coast Guard currently have limited infrastructure in place in the Arctic to support expanded ship and aircraft operations in the Arctic.[222]
- Expanded ship and aircraft operations in the Arctic may require altering ship and aircraft designs and operating methods.[223]
- Cooperation with other Arctic countries will be valuable in achieving defense and homeland security goals.

Navy in General

In a February 2009 journal article, the Oceanographer of the Navy stated:

> Competing claims dealing with the Arctic are often political in nature and have important implications. For example, in the summer of 2008 Canada announced that it would increase its military presence in the region, begin construction of a deep-water port on Baffin Island, establish a cold weather training base at Resolute Bay, and build six new ice-hardened ships to patrol the Northwest Passage. During the same period, Russia conducted strategic bomber flights over the area for the first time since the end of the Cold War....
>
> Preserving freedom of navigation in the region is an important tenet of U.S. policy. The NSR [Northern Sea Route], however, is a contested waterway, with Russian claims of sovereignty competing against U.S. and European Union insistence that it is an international strait available to all nations, subject to mutually recognized terms. Another potential [Arctic] transoceanic shipping route may be the Northwest Passage, which extends from the Atlantic through Baffin Bay and the Canadian Archipelago and into the Pacific by way of the Bering Strait. Canada claims sovereignty over the waters of the Canadian Archipelago, although the United States and the European Union claim that the Northwest Passage also constitutes an international strait which allows right of innocent passage....
>
> Aside from access and right of passage, the Navy and Coast Guard, in particular, must also be concerned with strategic choke points such as the Bering Strait, Canada's Queen Elizabeth Islands in the Northwest Passage, and Russia's Severnaya Zemlya and New Siberian Islands in the Northern Sea Route. These narrow passages offer some protection from persistent ice blockage, but they are also vulnerable to control or blockade by adversaries that would significantly disrupt potential commercial shipping and oil transport....
>
> U.S. naval interests will face new challenges in an increasingly ice-free Arctic with a strategic objective to understand potential threats to the United States from the maritime domain. As throughout the global commons, the U.S. Navy must be aware of activities that could be harmful to national security interests in a region that will, no doubt, see fewer barriers to access by potential adversaries in the future. National and homeland security interests pertinent to the U.S. Navy in the region would include early

> warning/missile defense; maritime presence and security; and freedom of navigation and over-flight....
>
> The region is primarily a maritime domain and the U.S. Navy of the future must be prepared to protect sea lines of communication supporting maritime commerce and other national interests—including national security—there. In addition to thinking through how we adjust our shipbuilding emphasis to support such operations, the Navy should also be thinking strategically about building the necessary infrastructure to provide logistic support for Arctic patrols, search and rescue capabilities, and shore-based support activities.
>
> To ensure complete maritime domain awareness in the region, and to provide our forces a competitive advantage, it will be necessary to have comprehensive knowledge of the physical environment. Data must be obtained by a suite of remote sensors (satellites, radars), autonomous sensors (data buoys, unmanned vehicles), and manned sensors (shipboard, coastal observing stations). Computer-based ocean and atmospheric models must be adjusted to the geophysical peculiarities of high latitudes. Communication lines for data exchange and reach-back processing at high-performance computing production centers must be robust and reliable. To ensure safety of navigation, we will also need to conduct more high-resolution bottom surveys and increase the scrutiny we place on sea ice conditions.
>
> The Navy relies on its international and interagency partners for assistance to ensure success of maritime domain awareness and maritime security missions. To meet the demands of national security in the changing northern environment, strengthening mechanisms for cooperation among the regional nations and U.S. agencies must remain a high priority. Like everywhere else in the world, sound national security in the Arctic will require strategic access, military mobility, safe navigation, unimpeded maritime transportation, improved homeland security, and responsible, sustainable use of ocean and coastal resources. International and interagency agreements and partnerships are vital to incorporating these essential elements into a viable national security policy and will be critical for resolving future naval challenges of a changing Arctic.[224]

In a spring 2010 journal article, the Oceanographer of the Navy (the successor to the officer who authored the above journal article) and a coauthor stated that "Navy surface vessels are able to operate up to the marginal ice zone but will require ice-strengthening to operate in higher ice conditions; Navy aircraft are capable of operating in the Arctic, but the lack of divert fields limits their duration and range."[225] The article also stated:

> While the Navy has a rich history in the Arctic, several challenges must be met to ensure successful operations in the future. These include the lack of support infrastructure and logistics support, environmental hazards such as drifting sea ice and icing on exposed surfaces, and communications difficulties. Antiquated nautical charts, drifting ice, low visibility, and the paucity of electronic and visual navigation aids hinder safety of navigation. A lack of coastal installations also contributes to the difficulty of search and rescue (SAR) operations. The only American-owned deepwater port near the Arctic basin is Dutch Harbor, in the Aleutian Islands.[226]

A March 2010 press report stated that

> a sustained [Navy] presence [in the Arctic] requires a lot of answers and training in the near future. Specifically, the harsh environment will require significant modifications in ship and satellite architecture.

"Where do we keep our fuel?" [Admiral Gary Roughead] the CNO [Chief of Naval Operations] said. "The way that we design our heating and ventilation systems are just not designed to keep up with ambient temperatures that are as cold as it gets up there. You have difficulties sometimes in how the more traditional outside equipment is maintained, and when you have freezing rain and things like that, how do you keep it clear? There are a lot of ship designs and ship alterations aspects.... What sort of communication architectures do we have to have? I really believe now is the time to start thinking about that because it means changes to ship design, which means we have to start laying in. It means perhaps some new overhead satellite architectures, and those are extraordinarily costly."[227]

On June 16, 2011, in remarks at a seminar on the Arctic, Admiral Gary Roughead, the Chief of Naval Operations, stated:

Well in my mind, there is a phenomenal event taking place on the planet today, and that is what I call the opening of the Fifth Ocean; that's the Arctic Ocean. We haven't had an ocean open on this planet since the end of the Ice Age. So if this is not a significant change that requires new, and I would submit, brave thinking on the topic, I don't know what other sort of physical event could produce that.

We in the Navy a couple of years ago began to look at how we as a Navy serve the nation.... So as we began to look at the planet, we realized that there were significant changes taking place; opening of the Fifth Ocean; physical changes; changes in population as they compressed into the littoral areas.

And so we established, I established something that I call Task Force Climate Change. Primarily focused on the Arctic because of the massive changes that are taking place up there, but it also takes into account what is happening on the rest of the planet and where we as a Navy think we have to be in the future. And then from that Task Force Climate Change, we developed what we call the Arctic road map....

The way we see things taking place is the first of moves, which will largely be in the area of resources. Fishing stocks will move with the water temperature, they will start to drift farther north, that will take fishing fleets farther north and there is a set of responsibilities that we have with other nations on making sure that that activity is taking place lawfully.

Then we believe the next step will be the extraction of resources. So what do we have to be prepared to do as a Navy, as a military particularly aligned closely with the Coast Guard on this.

And then in about 20 years time, 25 years time, the Arctic becomes a profitable sea route from Asia to Europe over the top of the planet. If you look at some of the estimates from shippers, some may actually be here. That is about a million dollars a trip someone saves. That is not insignificant. And so it will become a busy place as transportation starts to take place. That is how we envision the changes that are taking place, how we are looking at them, and the road map that we put in place is based on the best science as we know today. And we are committed to providing the resources the nation needs from the naval perspective to meet our needs.[228]

Four days later, in remarks to a symposium on the impacts on naval and maritime operations of an ice-diminishing Arctic, Admiral Roughead stated:

Since this forum began in 2001, our Navy has benefitted from the development and the discussion of the Arctic Symposium, as it has improved our appreciation greatly for an ice-diminished Arctic and the security implications that will follow from that physical change that will take place.

This broader examination is something that we've endeavored to focus on systematically since we stood up our Task Force Climate Change in May 2009, and issued our Arctic Roadmap that same year. And rather than all the answers, what our efforts have given us is an appreciation of just how dynamic the study of the region is and surely will be in the future....

... at a fundamental level, the trends point undeniably towards a new venue of operations and responsibility for our global Navy – for preserving American interests in free and fair access there – and in light of this we remain committed to preparing exhaustively for the challenges and especially for the opportunities that are going to exist in an ice-diminishing Arctic.

The U.S. Navy's interests in the Arctic are not new, of course. We have many decades of experience with exploration and, indeed, episodic operations in the waters of the Arctic Circle....

But never has our interest encountered the confluence of trends, as projected by the U.S. Geological Survey in 2008 and the National Research Council this past March, that promises to change the Arctic so pervasively, and in so doing affect the global environment for which we plan and program our future fleet....

In projecting the impact of climate change in the northern latitudes, however, I'm reminded of what Dr. Lubchenco observed just this past March, when she said, "what happens in the Arctic does not stay there." The trends we discuss here, in a similar timeframe, promise more disruption and disorder in a world whose population is growing rapidly, and moving to megacities on or near the coasts of almost every continent. The prospects of sea level rise, for some megacities, or the coral islands of the Maldives, are similarly daunting. We also have to consider the likely frictions that arise as fishing stocks migrate with changing sea temperatures, and the very real possibility that conflicts in the future will be fought over access to dwindling natural supplies of fresh water.

It is because of these projections that our Navy is preparing for increased demand, both in the region - where we will maintain our access and uphold the freedom of navigation as a global good - and beyond, where we expect developments to expose the costs of our national reluctance on the Law of the Sea convention and to test our present understanding of customary legal guarantees to the very freedoms behind our global operations today. We are considering the technical requirements for polar operations to support our strategic objective of a safe, stable, and secure Arctic region where our national interests are safeguarded – namely, how and when to build forces capable and competent for the harsh northern climes.

We also remain well aware of how important maritime partnerships will be in addressing capacity concerns as we seek to reinforce universal values in the global commons. Some speak of a changing Arctic in terms of either a race or a zero-sum game. As far as the U.S. Navy is concerned, it is neither. Rather, we see it as an opportunity to extend to the 'Fifth Ocean' the principles that have benefitted all peaceful nations in the other four. Secretary of State Clinton's attendance at the meeting of Arctic Council ministers last month in Nuuk, Greenland, signals the United States level of interest in the region, as well as our belief in the power of international cooperation to advance mutual interests in the free use of the maritime domain. It was also inherently practical in that it forged the much-needed agreement among Arctic states to improve our coordination on search and rescue.

As we stated in our Arctic Roadmap, the Navy will continue to pursue cooperative relationships around such areas of common interest, whether mission enablers like Maritime Domain Awareness, core naval capabilities such as humanitarian assistance and disaster response, or intra-governmental relationships like defense support to civil authorities. Exercises such as the Canadian Navy's 'Operation Nanook,' which we observed in 2009 and had the privilege of participating in last August alongside our Coast Guard counterparts, enhance our preparedness in several of these areas, and lay the basis from which we might extend the same maritime partnership and interoperability we have come to value so highly in our global operations. Joint exercises such as the U.S. Pacific Command's 'Northern Edge' – which started just outside of the Arctic Circle one week ago today and will continue throughout this week – ensure that we as a Navy remain ready to partner at the high end of operations as required.[229]

An April 28, 2012, press report states:

As global warming opens the Arctic Ocean to commercial and industrial traffic, the U.S. Navy is pushing to catch up with Russia, Canada and even Denmark in its Arctic ability. If a crisis were to happen now, the Navy lacks the ability to act in the Arctic without the help of one of those countries or the Coast Guard.

Last year, the Navy asked the War Gaming Department of the U.S. Naval War College to find out what the Navy needs for sustained operations in the Arctic.

In the resulting 2011 Fleet Arctic Operations Game, the Navy learned how big its Arctic shortcomings are. As a force, the Navy lacks everything from bases and Arctic-capable ships to reliable communications and cold-weather clothing....

The game's conclusions: the Navy is not adequately prepared to conduct long-term maritime Arctic operations; Arctic weather conditions increase the risk of failure; and most critically, to operate in the Arctic, the Navy will need to lean on the U.S. Coast Guard, countries like Russia or Canada, or tribal and industrial partners.

To sustain operations in the Arctic, the Navy needs ice capable equipment, accurate and timely environmental data, personnel trained to operate in extreme weather, and better communications systems. Much of the environmental data will come from other Arctic nations....

Navy officials understand the need to conduct exercises in the Arctic so they can get ready for the real thing, but they don't have a strategy.

"We are the only Arctic nation without an Arctic strategy," said U.S. Navy Cmdr. Blake McBride, Arctic Affairs officer for Task Force Climate Change. "The Coast Guard and Department of Defense are working on a strategy to help answer the issue, and advocate for capabilities."

Aside from signing National Security Presidential Directive 66, which requires the U.S. to have a presence in the Arctic, the Arctic hasn't been a priority for the U.S. government, largely because there isn't an immediate military threat.

"It's becoming a higher priority, but we don't make our own priorities," McBride said. "We don't foresee a military threat in the Arctic, but it doesn't mean we will not need to be able to operate there."

The Navy's future plans to conduct operations in the Arctic largely depend on the budget.

"It's all about the money," McBride said. "If you don't have the budget or funds to invest in manpower and equipment then you don't have anything."[230]

2009 Navy Arctic Roadmap

The Navy examined issues and concerns relating to climate change at a May 15, 2009, meeting of the Chief of Naval Operations (CNO) Executive Board. Following this meeting, the Navy decided to establish a Navy group led by the Oceanographer of the Navy called Task Force Climate Change (TFCC), and to develop Navy roadmaps first for the Arctic, and later for more general responses to global climate change.

The Navy issued its Arctic roadmap on November 10, 2009.[231] The document, which is dated October 2009 and co-sponsored by TFCC and the Oceanographer of the Navy,[232] is intended to guide the service's activities regarding the Arctic for the period FY2010-FY2014. The November 10 cover memo states that the document is to remain in effect until the completion of the report on the FY2014 Quadrennial Defense Review (QDR), at which time it will be reviewed and revised to incorporate guidance from that QDR.

The roadmap states that "significant action items" for FY2010 for the Navy included the following:

- Conduct assessments of fleet readiness and mission requirements in the Arctic region.
- Develop Navy strategic objectives in the Arctic region.
- Continue partnership-building activities with stakeholders in the region and conduct a limited objective experiment (or LOE—a kind of exercise) for the Arctic.
- Continue monitoring the Air Force's Polar Military Satellite Communications (MILSTACOM) program.
- Advocate for U.S. accession to the U.N. Convention on the Law of the Sea.
- Develop a Navy position regarding combatant commander authorities and responsibilities for the Arctic.
- Conduct research and development on a next-generation environmental prediction capability that is applicable to the Arctic.

The roadmap states that "significant action items" for FY2011-FY2012 for the Navy include the following:

- Initiate assessments of required Navy Arctic capabilities.
- Develop recommendations to address Arctic requirements for program proposals in the Navy's Program Objective Memorandum (or POM—a document that guides the development of a Navy budget) for FY2014 (POM-14).
- Continue biennial Navy participation in Arctic exercises, including ICEX-11, ICEX-13, Arctic Edge, and Arctic Care.
- Formalize new cooperative relationships that increase Navy experience and competency in search and rescue (SAR), maritime domain awareness (MDA), and humanitarian assistance and disaster response (HADR) in the Arctic, and defense support of civil authorities (DSCA) in Alaska.

The roadmap states that "significant action items" for FY2013-FY2014 for the Navy include the following:

- Execute Navy POM-14 budget initiatives that address Arctic requirements.
- Initiate combined (i.e., multi-lateral) and bilateral activities which support safety, security, and stability in the region.[233]

A January 2011 press report stated that

> Two studies that will define the new assets the Navy will need to operate in the Arctic and what those operations will look like are "nearing completion," according to

Rear Adm. David Titley, the oceanographer of the Navy, who spoke with Inside the Navy in a Jan. 18 interview.

The studies, called the Capabilities Based Assessment and the Mission Analysis, were both called for in the Navy's 2009 Arctic Roadmap, a strategic document meant to guide the service's efforts in the Arctic.

The capabilities assessment will undergo a flag officer review within the next month, according to Titley.[234]

Another January 2011 press report stated:

The Navy has decided to re-establish the Office of Naval Research's [ONR's] Arctic program as part of an increasing focus on the region, according to Navy officials....

In 2009, the Navy's Arctic Roadmap laid out the service's strategy for the region, stating that by the fourth quarter of fiscal year 2011 the Navy would "evaluate the re-establishment of ONR's High Latitude Program" to study the Arctic.

In August 2011, the Navy released an Arctic environment assessment and outlook report.[235] The report states that it

addresses Action Item 5.7 of the U.S. Navy Arctic Roadmap - to produce an Arctic Environmental Assessment and Outlook Report. This first biennial report provides a comprehensive assessment of the state of the Arctic environment, including the oceanography, hydrography, meteorology, fisheries, ice-extent, and climatic trends. This is important because the IPCC [reporting] refresh rate is too long to meet the budget POM [Program Objective Memorandum] cycle,[236] so this assessment will periodically synthesize existing scientific reports to inform POMs, specifically POM-[FY20]14; this allows the Navy's decisions to be based on sound science, and not use one source only, but a consensus of accepted sources....

As the Arctic environment continues to change and human activity increases, the U.S. Navy must be prepared to operate in this region. It is important to note that even though the Arctic is opening up, it will continue to be a harsh and challenging environment for the foreseeable future due to hazardous sea ice, freezing temperatures and extreme weather. Although the Navy submarine fleet has decades of experience operating in the Arctic, the surface fleet, air assets, and U.S. Marine Corps ground troops have limited experience there. The Navy must now consider the Arctic in terms of future policy, strategy, force structure, and investments.[237]

An April 28, 2012, press report states:

Navy officials have done the work called for in phase one and two of the [Arctic] roadmap, which largely consisted of developing research, assessing fleet readiness, completing capabilities-based assessments like the Fleet Arctic Operations Game, and formalizing cooperative agreements.

The biggest hurdle comes in the next phase, which calls for funding equipment and Arctic training. Navy officials say they are drafting a budget request to address those items.[238]

September 2010 GAO Report on Coast Guard's Arctic Requirements

A September 2010 GAO report on the Coast Guard's efforts to identify Arctic requirements in general stated:

The Coast Guard has taken specific action to identify Arctic requirements and gaps while also collecting relevant information from routine operations. The High Latitude Study is the centerpiece of the agency's efforts to determine its Arctic requirements. The Coast Guard has also established temporary operating locations in the Arctic and conducted biweekly Arctic overflights to obtain more information on the Arctic operating environment. In addition, information gathered during the Coast Guard's routine missions—ice breaking, search and rescue, and others—also informs requirements. The agency's preliminary efforts to identify its Arctic requirements generally align with key practices for agencies defining missions and desired outcomes.

The Coast Guard faces Arctic challenges including limited information, minimal assets and infrastructure, personnel issues, and difficult planning and funding decisions, but is taking initial steps to address these challenges. Specifically, the Coast Guard does not currently have Arctic maritime domain awareness—a full understanding of variables that could affect the security, safety, economy, or environment in the Arctic—but is acquiring additional Arctic vessel tracking data, among other things, to address this issue. In addition, the Coast Guard's Arctic assets and infrastructure are limited and not suitable for the harsh environment, but the agency is testing equipment and using alternative options to mitigate gaps. Finally, the Coast Guard faces uncertainty over the timing of predicted environmental changes in the Arctic, as well as over future funding streams. To address these challenges the Coast Guard obtains scientific data on Arctic climate change and is studying its Arctic resource requirements to support potential future funding needs.[239]

Coast Guard High Latitude Study Provided to Congress in July 2011

In July 2011, the Coast Guard provided to Congress a study on the Coast Guard's missions and capabilities for operations in high-latitude (i.e., polar) areas. The study, commonly known as the High Latitude Study, is dated July 2010 on its cover. The High Latitude Study concluded the following:

> [The study] concludes that future [Coast Guard] capability and capacity gaps will significantly impact four [Coast Guard] mission areas in the Arctic: Defense Readiness, Ice Operations, Marine Environmental Protection, and Ports, Waterways, and Coastal Security. These mission areas address the protection of important national interests in a geographic area where other nations are actively pursuing their own national goals. U.S. national policy and laws define the requirements to assert the nation's jurisdiction over its territory and interests; to ensure the security of its people and critical infrastructure; to participate fully in the collection of scientific knowledge; to support commercial enterprises with public utility; and to ensure that the Arctic environment is not degraded by increased human activity.
>
> The Coast Guard's ability to support Defense Readiness mission requirements in the Arctic is closely linked to DoD responsibilities. The Coast Guard presently possesses the only surface vessels capable of operating in ice-covered and ice-diminished waters. The Coast Guard supports (1) DoD missions such as the resupply of Thule Air Base in Greenland and logistics support (backup) for McMurdo Station in Antarctica and (2) Department of State (DoS) directed Freedom of Navigation Operations. These unique Coast Guard capabilities have been noted by the Joint Chiefs of Staff, the Navy's Task Force Climate Change, and the recently issued Naval Operations Concept 2010.
>
> The common and dominant contributor to these significant mission impacts is the gap in polar icebreaking capability....[240]

Other capability gaps contributing to the impact on Coast Guard ability to carry out its missions in the Arctic include:

- Communications System Capability – Continuous coverage along Alaska's West Coast, the Bering Strait, and throughout the North Slope is required for exchanging voice and data communications with Coast Guard units and other government and commercial platforms offshore.
- Forward Operating Locations - No suitable facilities currently exist on the North Slope or near the Bering Strait with facilities sufficient to support extended aircraft servicing and maintenance. Aircraft must travel long distances and expend significant time transiting to and from adequate facilities. This gap reduces on-scene presence and capability to support sustained operations in the region.
- Environmental response in ice-covered waters - The technology and procedures for assessment and mitigation measures for oil spills in ice-covered waters are not fully developed or tested.

Capability gaps in the Arctic region have moderate impacts on [the Coast Guard's] Aids to Navigation (AtoN), Search and Rescue (SAR), and Other Law Enforcement (OLE) missions. Both AtoN and SAR involve the safety of mariners and will gain more importance not only as commerce and tourism cause an increase in maritime traffic, but as U.S. citizens in northern Alaska face more unpredictable conditions. Performance of OLE will be increasingly necessary to ensure the integrity of U.S. living marine resources from outside pressures....

In addition to the assessment of polar icebreaking needs, the Arctic mission analysis examined a set of theoretical mixes (force packages) of Coast Guard assets consisting of icebreakers, their embarked helicopters, and deployment alternatives using aviation forward operating locations in Arctic Alaska....

All [six] of the force mixes [considered in the study] add assets to the existing Coast Guard Alaska Patrol consisting of (1) a high-endurance cutter (not an icebreaker) deployed in the Bering Sea carrying a short range recovery helicopter, and (2) medium range recovery helicopters located at Kodiak in the Gulf of Alaska, and seasonally deployed to locations in Cold Bay and St. Paul Island....

These force packages and associated risk assessment provide a framework for acquisition planning as the Coast Guard implements a strategy for closing the capability gaps. By first recapitalizing the aging icebreakers, the Coast Guard provides a foundation for buildout of these force mixes. In addition to the cost of the icebreakers, the force packages require investment in forward operating locations and in medium range helicopters. The mission analysis reports developed rough order-of-magnitude cost estimates for forward operating locations at approximately $36M [million] each and for helicopters at $9M each....

The analysis shows that the current Coast Guard deployment posture is not capable of effective response in northern Alaska and that response may be improved through a mix of deployed cutters, aircraft, and supporting infrastructure including forward operating locations and communications/navigation systems.[241]

U.S.-Canadian Military Cooperation

In December 2009, it was reported that "U.S. and Canadian defense officials are studying emerging gaps in their awareness of Arctic activities, seeking to boost North American Aerospace Defense [NORAD] Command's maritime-warning mission and crafting a new threat assessment for the region." The effort would reportedly involve both NORAD and a Canadian-American advisory board called the Permanent Joint Board on Defense.[242]

In May 2010, it was reported that "American and Canadian defense officials are bolstering collaboration on military exercises, investment plans and technology development related to the Arctic." The report stated that "U.S. and Canadian intelligence officials have created a classified joint "utilization" assessment for the Arctic looking out to 2020, which will be continuously updated." It also stated that "American and Canadian defense officials have just started implementing a new five-year work plan that contains specific initiatives and requires concrete deliverables, Stockton said, noting the Arctic received special attention in the plan. The plan also covers defense critical infrastructure protection, defense support to civil agencies and defense cooperation in the Americas."[243]

In December 2012, the commander of NORAD and USNORTHCOM and the commander of the Canadian Joint Operations Command signed two documents on Arctic cooperation. A December 11, 2012, statement issued by NORAD and USNORTHCOM on the signing of the documents stated in part:

> Army Gen. Charles Jacoby, Jr., commander of NORAD and U.S. Northern Command, and Lt.-Gen. Stuart Beare, Canadian Joint Operations Command commander, signed two significant documents today—the Tri Command Framework for Arctic Cooperation and the Tri-Command Training and Exercise Statement of Intent—during the 230th meeting of the Canada-U.S. Permanent Joint Board on Defense in Colorado Springs, Colo.
>
> The Tri-Command Framework for Arctic Cooperation acknowledges the Arctic is not a region of conflict and the Canadian and U.S. militaries will support other departments and agencies in response to threats and hazards in the region when requested and directed. In that context, the goal of the Framework is to promote enhanced military cooperation in the Arctic and identify specific areas of potential Tri-Command cooperation in the preparation for and conduct of safety, security and defense operations. It strengthens an already unique and mature partnership where coordination and cooperation occurs on a regular basis. The TriCommand Framework for Arctic Cooperation document is not a plan but rather outlines a process that supports the identification of opportunities for potential cooperation in the Arctic. Areas that continue to be improved, particularly in the Arctic, include planning, domain awareness, information-sharing, training and exercises, operations, capability development, and science and technology....
>
> The second document, the Tri Command Training and Exercise Statement of Intent, is aimed at enhancing joint and combined readiness in support of safety, security and defence missions through combined training and exercises and reinforcing partnerships and collaboration among the Commands. The Tri-Command Training and Exercise Directive will help to ensure a timely and coordinated response to safety, security and defence challenges to North America.
>
> The signing of the Tri Command Framework for Arctic Cooperation and the Tri-Command Training and Exercise Statement of Intent follows on a Tri Command Strategy that spells out a series of shared tasks designed to strengthen working relationships with all Tri-Command defense and security partners.[244]

January 2009 Arctic Policy Directive

As noted earlier (see "January 2009 Arctic Policy Directive" in "Background"), the Obama Administration is currently operating under the January 2009 Arctic region policy directive (NSPD 66/HSPD 25) issued by the George W. Bush Administration.[245] Potential oversight questions include but not are limited to the following:

- Is NSPD 66/HSPD 25 a suitable statement of U.S. policy for the Arctic region?[246]
- Although the Obama Administration is currently operating under NSPD 66/HSPD 25, does the Obama Administration fully agree with all parts of it? If not, with which parts does it not fully agree?
- Does the Obama Administration intend to eventually conduct a review of NSDP 66/HSPD 25? If so, what is the Administration's schedule for conducting and releasing the results of that review?

Appendix A. Arctic Research and Policy Act (ARPA) of 1984 (Title I of P.L. 98-373)

The text of the Arctic Research and Policy Act (ARPA) of 1984 (Title I of P.L. 98-373 of July 31, 1984)[247] is as follows:

TITLE I – ARCTIC RESEARCH AND POLICY

SHORT TITLE

SEC. 101. This title may be cited as the "Arctic Research and Policy Act of 1984".

FINDINGS AND PURPOSES

SEC. 102. (a) The Congress finds and declares that-

(1) the Arctic, onshore and offshore, contains vital energy resources that can reduce the Nation's dependence on foreign oil and improve the national balance of payments;

(2) as the Nation's only common border with the Soviet Union, the Arctic is critical to national defense;

(3) the renewable resources of the Arctic, specifically fish and other seafood, represent one of the Nation's greatest commercial assets;

(4) Arctic conditions directly affect global weather patterns and must be understood in order to promote better agricultural management throughout the United States;

(5) industrial pollution not originating in the Arctic region collects in the polar air mass, has the potential to disrupt global weather patterns, and must be controlled through international cooperation and consultation;

(6) the Arctic is a natural laboratory for research into human health and adaptation, physical and psychological, to climates of extreme cold and isolation and may provide information crucial for future defense needs;

(7) atmospheric conditions peculiar to the Arctic make the Arctic a unique testing ground for research into high latitude communications, which is likely to be crucial for future defense needs;

(8) Arctic marine technology is critical to cost-effective recovery and transportation of energy resources and to the national defense;

(9) the United States has important security, economic, and environmental interests in developing and maintaining a fleet of icebreaking vessels capable of operating effectively in the heavy ice regions of the Arctic;

(10) most Arctic-rim countries, particularly the Soviet Union, possess Arctic technologies far more advanced than those currently available in the United States;

(11) Federal Arctic research is fragmented and uncoordinated at the present time, leading to the neglect of certain areas of research and to unnecessary duplication of effort in other areas of research;

(12) improved logistical coordination and support for Arctic research and better dissemination of research data and information is necessary to increase the efficiency and utility of national Arctic research efforts;

(13) a comprehensive national policy and program plan to organize and fund currently neglected scientific research with respect to the Arctic is necessary to fulfill national objectives in Arctic research;

(14) the Federal Government, in cooperation with State and local governments, should focus its efforts on the collection and characterization of basic data related to biological, materials, geophysical, social, and behavioral phenomena in the Arctic;

(15) research into the long-range health, environmental, and social effects of development in the Arctic is necessary to mitigate the adverse consequences of that development to the land and its residents;

(16) Arctic research expands knowledge of the Arctic, which can enhance the lives of Arctic residents, increase opportunities for international cooperation among Arctic-rim countries, and facilitate the formulation of national policy for the Arctic; and

(17) the Alaskan Arctic provides an essential habitat for marine mammals, migratory waterfowl, and other forms of wildlife which are important to the Nation and which are essential to Arctic residents.

(b) The purposes of this title are-

(1) to establish national policy, priorities, and goals and to provide a Federal program plan for basic and applied scientific research with respect to the Arctic, including natural resources and materials, physical, biological and health sciences, and social and behavioral sciences;

(2) to establish an Arctic Research Commission to promote Arctic research and to recommend Arctic research policy;

(3) to designate the National Science Foundation as the lead agency responsible for implementing Arctic research policy; and

(4) to establish an Interagency Arctic Research Policy Committee to develop a national Arctic research policy and a five year plan to implement that policy.

ARCTIC RESEARCH COMMISSION

SEC. 103. (a) The President shall establish an Arctic Research Commission (hereafter referred to as the "Commission").

(b)(1) The Commission shall be composed of five members appointed by the President, with the Director of the National Science Foundation serving as a nonvoting, ex officio member. The members appointed by the President shall include-

(A) three members appointed from among individuals from academic or other research institutions with expertise in areas of research relating to the Arctic, including the physical, biological, health, environmental, social, and behavioral sciences;

(B) one member appointed from among indigenous residents of the Arctic who are representative of the needs and interests of Arctic residents and who live in areas directly affected by Arctic resource development; and

(C) one member appointed from among individuals familiar with the Arctic and representative of the needs and interests of private industry undertaking resource development in the Arctic.

(2) The President shall designate one of the appointed members of the Commission to be chairperson of the Commission.

(c)(1) Except as provided in paragraph (2) of this subsection, the term of office of each member of the Commission appointed under subsection (b)(1) shall be four years.

(2) Of the members of the Commission originally appointed under subsection (b)(1)-

(A) one shall be appointed for a term of two years;

(B) two shall be appointed for a term of three years; and

(C) two shall be appointed for a term of four years.

(3) Any vacancy occurring in the membership of the Commission shall be filled, after notice of the vacancy is published in the Federal Register, in the manner provided by the preceding provisions of this section, for the remainder of the unexpired term.

(4) A member may serve after the expiration of the member's term of office until the President appoints a successor.

(5) A member may serve consecutive terms beyond the member's original appointment.

(d)(1) Members of the Commission may be allowed travel expenses, including per diem in lieu of subsistence, as authorized by section 5703 of title 5, United States Code.

A member of the Commission not presently employed for compensation shall be compensated at a rate equal to the daily equivalent of the rate for GS-16 of the General Schedule under section 5332 of title 5, United States Code, for each day the member is engaged in the actual performance of his duties as a member of the Commission, not to exceed 90 days of service each year. Except for the purposes of chapter 81 of title 5 (relating to compensation for work injuries) and chapter 171 of title 28 (relating to tort claims), a member of the Commission shall not be considered an employee of the United States for any purpose.

(2) The Commission shall meet at the call of its Chairman or a majority of its members.

(3) Each Federal agency referred to in section 107(b) may designate a representative to participate as an observer with the Commission.

These representatives shall report to and advise the Commission on the activities relating to Arctic research of their agencies.

(4) The Commission shall conduct at least one public meeting in the State of Alaska annually.

DUTIES OF COMMISSION

SEC. 104. (a) The Commission shall-

(1) develop and recommend an integrated national Arctic research policy;

(2) in cooperation with the Interagency Arctic Research Policy Committee established under section 107, assist in establishing a national Arctic research program plan to implement the Arctic research policy;

(3) facilitate cooperation between the Federal Government and State and local governments with respect to Arctic research;

(4) review Federal research programs in the Arctic and suggest improvements in coordination among programs;

(5) recommend methods to improve logistical planning and support for Arctic research as may be appropriate and in accordance with the findings and purposes of this title;

(6) suggest methods for improving efficient sharing and dissemination of data and information on the Arctic among interested public and private institutions;

(7) offer other recommendations and advice to the Interagency Committee established under section 107 as it may find appropriate; and

(8) cooperate with the Governor of the State of Alaska and with agencies and organizations of that State which the Governor may designate with respect to the formulation of Arctic research policy.

(b) Not later than January 31 of each year, the Commission shall-

(1) publish a statement of goals and objectives with respect to Arctic research to guide the Interagency Committee established under section 107 in the performance of its duties; and

(2) submit to the President and to the Congress a report describing the activities and accomplishments of the Commission during the immediately preceding fiscal year.

COOPERATION WITH THE COMMISSION

SEC. 105. (a)(1) The Commission may acquire from the head of any Federal agency unclassified data, reports, and other nonproprietary information with respect to Arctic research in the possession of the agency which the Commission considers useful in the discharge of its duties.

(2) Each agency shall cooperate with the Commission and furnish all data, reports, and other information requested by the Commission to the extent permitted by law; except that no agency need furnish any information which it is permitted to withhold under section 552 of title 5, United States Code.

(b) With the consent of the appropriate agency head, the Commission may utilize the facilities and services of any Federal agency to the extent that the facilities and services are needed for the establishment and development of an Arctic research policy, upon reimbursement to be agreed upon by the Commission and the agency head and taking every feasible step to avoid duplication of effort.

(c) All Federal agencies shall consult with the Commission before undertaking major Federal actions relating to Arctic research.

ADMINISTRATION OF THE COMMISSION

SEC. 106. The Commission may-

(1) in accordance with the civil service laws and subchapter III of chapter 53 of title 5, United States Code, appoint and fix the compensation of an Executive Director and necessary additional staff personnel, but not to exceed a total of seven compensated personnel;

(2) procure temporary and intermittent services as authorized by section 3109 of title 5, United States Code;

(3) enter into contracts and procure supplies, services, and personal property; and

(4) enter into agreements with the General Services Administration for the procurement of necessary financial and administrative services, for which payment shall be made by reimbursement from funds of the Commission in amounts to be agreed upon by the Commission and the Administrator of the General Services Administration.

LEAD AGENCY AND INTERAGENCY ARCTIC RESEARCH POLICY COMMITTEE

SEC. 107. (a) The National Science Foundation is designated as the lead agency responsible for implementing Arctic research policy, and the Director of the National Science Foundation shall insure that the requirements of section 108 are fulfilled.

(b)(1) The President shall establish an Interagency Arctic Research Policy Committee (hereinafter referred to as the "Interagency Committee").

(2) The Interagency Committee shall be composed of representatives of the following Federal agencies or offices:

(A) the National Science Foundation;

(B) the Department of Commerce;

(C) the Department of Defense;

(D) the Department of Energy;

(E) the Department of the Interior;

(F) the Department of State;

(G)the Department of Transportation;

(H) the Department of Health and Human Services;

(I) the National Aeronautics and Space Administration;

(J) the Environmental Protection Agency; and

(K) any other agency or office deemed appropriate.

(3) The representative of the National Science Foundation shall serve as the Chairperson of the Interagency Committee.

DUTIES OF THE INTERAGENCY COMMITTEE

SEC. 108. (a) The Interagency Committee shall-

(1) survey Arctic research conducted by Federal, State, and local agencies, universities, and other public and private institutions to help determine priorities for future Arctic research, including natural resources and materials, physical and biological sciences, and social and behavioral sciences;

(2) work with the Commission to develop and establish an integrated national Arctic research policy that will guide Federal agencies in developing and implementing their research programs in the Arctic;

(3) consult with the Commission on-

(A) the development of the national Arctic research policy and the 5-year plan implementing the policy;

(B) Arctic research programs of Federal agencies;

(C) recommendations of the Commission on future Arctic research; and

(D) guidelines for Federal agencies for awarding and administering Arctic research grants;

(4) develop a 5-year plan to implement the national policy, as provided for in section 109;

(5) provide the necessary coordination, data, and assistance for the preparation of a single integrated, coherent, and multiagency budget request for Arctic research as provided for in section 110;

(6) facilitate cooperation between the Federal Government and State and local governments in Arctic research, and recommend the undertaking of neglected areas of research in accordance with the findings and purposes of this title;

(7) coordinate and promote cooperative Arctic scientific research programs with other nations, subject to the foreign policy guidance of the Secretary of State;

(8) cooperate with the Governor of the State of Alaska in fulfilling its responsibilities under this title;

(9) promote Federal interagency coordination of all Arctic research activities, including- (A) logistical planning and coordination; and

(B) the sharing of data and information associated with Arctic research, subject to section 552 of title 5, United States Code; and

(10) provide public notice of its meetings and an opportunity for the public to participate in the development and implementation of national Arctic research policy.

(b) Not later than January 31, 1986, and biennially thereafter, the Interagency Committee shall submit to the Congress through the President, a brief, concise report containing-

(1) a statement of the activities and accomplishments of the Interagency Committee since its last report; and

(2) a description of the activities of the Commission, detailing with particularity the recommendations of the Commission with respect to Federal activities in Arctic research.

5-YEAR ARCTIC RESEARCH PLAN

SEC. 109. (a) The Interagency Committee, in consultation with the Commission, the Governor of the State of Alaska, the residents of the Arctic, the private sector, and public interest groups, shall prepare a comprehensive 5-year program plan (hereinafter referred to as the "Plan") for the overall Federal effort in Arctic research. The Plan shall be prepared and submitted to the President for transmittal to the Congress within one year after the enactment of this Act and shall be revised biennially thereafter.

(b) The Plan shall contain but need not be limited to the following elements:

(1) an assessment of national needs and problems regarding the Arctic and the research necessary to address those needs or problems;

(2) a statement of the goals and objectives of the Interagency Committee for national Arctic research;

(3) a detailed listing of all existing Federal programs relating to Arctic research, including the existing goals, funding levels for each of the 5 following fiscal years, and the funds currently being expended to conduct the programs;

(4) recommendations for necessary program changes and other proposals to meet the requirements of the policy and goals as set forth by the Commission and in the Plan as currently in effect; and

(5) a description of the actions taken by the Interagency Committee to coordinate the budget review process in order to ensure interagency coordination and cooperation in (A) carrying out Federal Arctic research programs, and (B) eliminating unnecessary duplication of effort among these programs.

COORDINATION AND REVIEW OF BUDGET REQUESTS

SEC. 110. (a) The Office of Science and Technology Policy shall-

(1) review all agency and department budget requests related to the Arctic transmitted pursuant to section 108(a)(5), in accordance with the national Arctic research policy and the 5-year program under section 108(a)(2) and section 109, respectively; and

(2) consult closely with the Interagency Committee and the Commission to guide the Office of Science and Technology Policy's efforts.

(b)(1) The Office of Management and Budget shall consider all Federal agency requests for research related to the Arctic as one integrated, coherent, and multiagency request which shall be reviewed by the Office of Management and Budget prior to submission of the President's annual budget request for its adherence to the Plan. The Commission shall, after submission of the President's annual budget request, review the request and report to Congress on adherence to the Plan.

(2) The Office of Management and Budget shall seek to facilitate planning for the design, procurement, maintenance, deployment, and operations of icebreakers needed to provide a platform for Arctic research by allocating all funds necessary to support icebreaking operations, except for recurring incremental costs associated with specific projects, to the Coast Guard.

AUTHORIZATION OF APPROPRIATIONS; NEW SPENDING AUTHORITY

SEC. 111. (a) There are authorized to be appropriated such sums as may be necessary for carrying out this title.

(b) Any new spending authority (within the meaning of section 401 of the Congressional Budget Act of 1974) which is provided under this title shall be effective for any fiscal year only to such extent or in such amounts as may be provided in appropriation Acts.

DEFINITION

SEC. 112. As used in this title, the term "Arctic" means all United States and foreign territory north of the Arctic Circle and all United States territory north and west of the boundary formed by the Porcupine, Yukon, and Kuskokwim Rivers; all contiguous seas, including the Arctic Ocean and the Beaufort, Bering, and Chukchi Seas; and the Aleutian chain.

APPENDIX B. P.L. 101-609 OF 1990, AMENDING ARCTIC RESEARCH AND POLICY ACT (ARPA) OF 1984

The Arctic Research and Policy Act (ARPA) of 1984 (see Appendix A) was amended by P.L. 101-609 of November 16, 1990. The text of P.L. 101-609 is as follows:

SECTION 1. Except as specifically provided in this Act, whenever in this Act an amendment or repeal is expressed as an amendment to, or repeal of a provision, the reference shall be deemed to be made to the Arctic Research and Policy Act of 1984.

SEC. 2. Section 103(b)(1) (15 U.S.C. 4102(b)(1)) is amended—

(1) in the text above clause (A), by striking out `five' and inserting in lieu thereof `seven';

(2) in clause (A), by striking out `three' and inserting in lieu thereof `four'; and

(3) in clause (C), by striking out `one member' and inserting in lieu thereof `two members'.

SEC. 3. Section 103(d)(1) (15 U.S.C. 4102(d)(1)) is amended by striking out `GS-16' and inserting in lieu thereof `GS-18'.

SEC. 4. (a) Section 104(a) (15 U.S.C. 4102(a)) is amended—

(1) in paragraph (4), by striking out `suggest' and inserting in lieu thereof `recommend';

(2) in paragraph (6), by striking out `suggest' and inserting in lieu thereof `recommend';

(3) in paragraph (7), by striking out `and' at the end thereof;

(4) in paragraph (8), by striking out the period and inserting in lieu thereof a semicolon; and

(5) by adding at the end thereof the following new paragraphs:

(9) recommend to the Interagency Committee the means for developing international scientific cooperation in the Arctic; and

not later than January 31, 1991, and every 2 years thereafter, publish a statement of goals and objectives with respect to Arctic research to guide the Interagency Committee established under section 107 in the performance of its duties.'.

(b) Section 104(b) is amended to read as follows:

`(b) Not later than January 31 of each year, the Commission shall submit to the President and to the Congress a report describing the activities and accomplishments of the Commission during the immediately preceding fiscal year.'.

SEC. 5. Section 106 (15 U.S.C. 4105) is amended—

(1) in paragraph (3), by striking out `and' at the end thereof;

(2) in paragraph (4), by striking out the period at the end thereof and inserting in lieu thereof `; and'; and

(3) by adding at the end thereof the following new paragraph:

`(5) appoint, and accept without compensation the services of, scientists and engineering specialists to be advisors to the Commission. Each advisor may be allowed travel expenses, including per diem in lieu of subsistence, as authorized by section 5703 of title 5, United

States Code. Except for the purposes of chapter 81 of title 5 (relating to compensation for work injuries) and chapter 171 of title 28 (relating to tort claims) of the United States Code, an advisor appointed under this paragraph shall not be considered an employee of the United States for any purpose.'.

SEC. 6. Subsection (b)(2) of section 108 (15 U.S.C. 4107(b)(2)) is amended to read as follows:

`(2) a statement detailing with particularity the recommendations of the Commission with respect to Federal interagency activities in Arctic research and the disposition and responses to those recommendations.'.

APPENDIX C. JANUARY 2009 ARCTIC POLICY DIRECTIVE (NSPD 66/HSPD 25)

On January 12, 2009, the George W. Bush Administration released a presidential directive establishing a new U.S. policy for the Arctic region. The directive, dated January 9, 2009, was issued as National Security Presidential Directive 66/Homeland Security Presidential Directive 25 (NSPD 66/HSPD 25). The text of NSPD 66/HSPD 25 is as follows:

SUBJECT: Arctic Region Policy

I. PURPOSE

A. This directive establishes the policy of the United States with respect to the Arctic region and directs related implementation actions. This directive supersedes Presidential Decision Directive/NSC-26 (PDD-26; issued 1994) with respect to Arctic policy but not Antarctic policy; PDD-26 remains in effect for Antarctic policy only.

B. This directive shall be implemented in a manner consistent with the Constitution and laws of the United States, with the obligations of the United States under the treaties and other international agreements to which the United States is a party, and with customary international law as recognized by the United States, including with respect to the law of the sea.

II. BACKGROUND

A. The United States is an Arctic nation, with varied and compelling interests in that region. This directive takes into account several developments, including, among others:

1. Altered national policies on homeland security and defense;
2. The effects of climate change and increasing human activity in the Arctic region;
3. The establishment and ongoing work of the Arctic Council; and
4. A growing awareness that the Arctic region is both fragile and rich in resources.

III. POLICY

A. It is the policy of the United States to:

1. Meet national security and homeland security needs relevant to the Arctic region;
2. Protect the Arctic environment and conserve its biological resources;

3. Ensure that natural resource management and economic development in the region are environmentally sustainable;

4. Strengthen institutions for cooperation among the eight Arctic nations (the United States, Canada, Denmark, Finland, Iceland, Norway, the Russian Federation, and Sweden);

5. Involve the Arctic's indigenous communities in decisions that affect them; and

6. Enhance scientific monitoring and research into local, regional, and global environmental issues.

B. National Security and Homeland Security Interests in the Arctic

1. The United States has broad and fundamental national security interests in the Arctic region and is prepared to operate either independently or in conjunction with other states to safeguard these interests. These interests include such matters as missile defense and early warning; deployment of sea and air systems for strategic sealift, strategic deterrence, maritime presence, and maritime security operations; and ensuring freedom of navigation and overflight.

2. The United States also has fundamental homeland security interests in preventing terrorist attacks and mitigating those criminal or hostile acts that could increase the United States vulnerability to terrorism in the Arctic region.

3. The Arctic region is primarily a maritime domain; as such, existing policies and authorities relating to maritime areas continue to apply, including those relating to law enforcement.[1] Human activity in the Arctic region is increasing and is projected to increase further in coming years. This requires the United States to assert a more active and influential national presence to protect its Arctic interests and to project sea power throughout the region.

4. The United States exercises authority in accordance with lawful claims of United States sovereignty, sovereign rights, and jurisdiction in the Arctic region, including sovereignty within the territorial sea, sovereign rights and jurisdiction within the United States exclusive economic zone and on the continental shelf, and appropriate control in the United States contiguous zone.

5. Freedom of the seas is a top national priority. The Northwest Passage is a strait used for international navigation, and the Northern Sea Route includes straits used for international navigation; the regime of transit passage applies to passage through those straits. Preserving the rights and duties relating to navigation and overflight in the Arctic region supports our ability to exercise these rights throughout the world, including through strategic straits.

6. Implementation: In carrying out this policy as it relates to national security and homeland security interests in the Arctic, the Secretaries of State, Defense, and Homeland Security, in coordination with heads of other relevant executive departments and agencies, shall:

a. Develop greater capabilities and capacity, as necessary, to protect United States air, land, and sea borders in the Arctic region;

b. Increase Arctic maritime domain awareness in order to protect maritime commerce, critical infrastructure, and key resources;

c. Preserve the global mobility of United States military and civilian vessels and aircraft throughout the Arctic region;

d. Project a sovereign United States maritime presence in the Arctic in support of essential United States interests; and

e. Encourage the peaceful resolution of disputes in the Arctic region.

C. International Governance

1. The United States participates in a variety of fora, international organizations, and bilateral contacts that promote United States interests in the Arctic. These include the Arctic Council, the International Maritime Organization (IMO), wildlife conservation and management agreements, and many other mechanisms. As the Arctic changes and human activity in the region increases, the United States and other governments should consider, as appropriate, new international arrangements or enhancements to existing arrangements.

2. The Arctic Council has produced positive results for the United States by working within its limited mandate of environmental protection and sustainable development. Its subsidiary bodies, with help from many United States agencies, have developed and undertaken projects on a wide range of topics. The Council also provides a beneficial venue for interaction with indigenous groups. It is the position of the United States that the Arctic Council should remain a high-level forum devoted to issues within its current mandate and not be transformed into a formal international organization, particularly one with assessed contributions. The United States is nevertheless open to updating the structure of the Council, including consolidation of, or making operational changes to, its subsidiary bodies, to the extent such changes can clearly improve the Council's work and are consistent with the general mandate of the Council.

3. The geopolitical circumstances of the Arctic region differ sufficiently from those of the Antarctic region such that an "Arctic Treaty" of broad scope—along the lines of the Antarctic Treaty—is not appropriate or necessary.

4. The Senate should act favorably on U.S. accession to the U.N. Convention on the Law of the Sea promptly, to protect and advance U.S. interests, including with respect to the Arctic. Joining will serve the national security interests of the United States, including the maritime mobility of our Armed Forces worldwide. It will secure U.S. sovereign rights over extensive marine areas, including the valuable natural resources they contain. Accession will promote U.S. interests in the environmental health of the oceans. And it will give the United States a seat at the table when the rights that are vital to our interests are debated and interpreted.

5. Implementation: In carrying out this policy as it relates to international governance, the Secretary of State, in coordination with heads of other relevant executive departments and agencies, shall:

a. Continue to cooperate with other countries on Arctic issues through the United Nations (U.N.) and its specialized agencies, as well as through treaties such as the U.N. Framework Convention on Climate Change, the Convention on International Trade in Endangered Species of Wild Fauna and Flora, the Convention on Long Range Transboundary Air Pollution and its protocols, and the Montreal Protocol on Substances that Deplete the Ozone Layer;

b. Consider, as appropriate, new or enhanced international arrangements for the Arctic to address issues likely to arise from expected increases in human activity in that region, including shipping, local development and subsistence, exploitation of living marine resources, development of energy and other resources, and tourism;

c. Review Arctic Council policy recommendations developed within the ambit of the Council's scientific reviews and ensure the policy recommendations are subject to review by Arctic governments; and

d. Continue to seek advice and consent of the United States Senate to accede to the 1982 Law of the Sea Convention.

D. Extended Continental Shelf and Boundary Issues

1. Defining with certainty the area of the Arctic seabed and subsoil in which the United States may exercise its sovereign rights over natural resources such as oil, natural gas, methane hydrates, minerals, and living marine species is critical to our national interests in energy security, resource management, and environmental protection. The most effective way to achieve international recognition and legal certainty for our extended continental shelf is through the procedure available to States Parties to the U.N. Convention on the Law of the Sea.

2. The United States and Canada have an unresolved boundary in the Beaufort Sea. United States policy recognizes a boundary in this area based on equidistance. The United States recognizes that the boundary area may contain oil, natural gas, and other resources.

3. The United States and Russia are abiding by the terms of a maritime boundary treaty concluded in 1990, pending its entry into force. The United States is prepared to enter the agreement into force once ratified by the Russian Federation.

4. Implementation: In carrying out this policy as it relates to extended continental shelf and boundary issues, the Secretary of State, in coordination with heads of other relevant executive departments and agencies, shall:

a. Take all actions necessary to establish the outer limit of the continental shelf appertaining to the United States, in the Arctic and in other regions, to the fullest extent permitted under international law;

b. Consider the conservation and management of natural resources during the process of delimiting the extended continental shelf; and

c. Continue to urge the Russian Federation to ratify the 1990 United States-Russia maritime boundary agreement.

E. Promoting International Scientific Cooperation

1. Scientific research is vital for the promotion of United States interests in the Arctic region. Successful conduct of U.S. research in the Arctic region requires access throughout the Arctic Ocean and to terrestrial sites, as well as viable international mechanisms for sharing access to research platforms and timely exchange of samples, data, and analyses. Better coordination with the Russian Federation, facilitating access to its domain, is particularly important.

2. The United States promotes the sharing of Arctic research platforms with other countries in support of collaborative research that advances fundamental understanding of the Arctic region in general and potential Arctic change in particular. This could include collaboration with bodies such as the Nordic Council and the European Polar Consortium, as well as with individual nations.

3. Accurate prediction of future environmental and climate change on a regional basis, and the delivery of near real-time information to end-users, requires obtaining, analyzing, and disseminating accurate data from the entire Arctic region, including both paleoclimatic data

and observational data. The United States has made significant investments in the infrastructure needed to collect environmental data in the Arctic region, including the establishment of portions of an Arctic circumpolar observing network through a partnership among United States agencies, academic collaborators, and Arctic residents. The United States promotes active involvement of all Arctic nations in these efforts in order to advance scientific understanding that could provide the basis for assessing future impacts and proposed response strategies.

4. United States platforms capable of supporting forefront research in the Arctic Ocean, including portions expected to be ice-covered for the foreseeable future, as well as seasonally ice-free regions, should work with those of other nations through the establishment of an Arctic circumpolar observing network. All Arctic nations are members of the Group on Earth Observations partnership, which provides a framework for organizing an international approach to environmental observations in the region. In addition, the United States recognizes that academic and research institutions are vital partners in promoting and conducting Arctic research.

5. Implementation: In carrying out this policy as it relates to promoting scientific international cooperation, the Secretaries of State, the Interior, and Commerce and the Director of the National Science Foundation, in coordination with heads of other relevant executive departments and agencies, shall:

a. Continue to play a leadership role in research throughout the Arctic region;

b. Actively promote full and appropriate access by scientists to Arctic research sites through bilateral and multilateral measures and by other means;

c. Lead the effort to establish an effective Arctic circumpolar observing network with broad partnership from other relevant nations;

d. Promote regular meetings of Arctic science ministers or research council heads to share information concerning scientific research opportunities and to improve coordination of international Arctic research programs;

e. Work with the Interagency Arctic Research Policy Committee (IARPC) to promote research that is strategically linked to U.S. policies articulated in this directive, with input from the Arctic Research Commission; and

f. Strengthen partnerships with academic and research institutions and build upon the relationships these institutions have with their counterparts in other nations.

F. Maritime Transportation in the Arctic Region

1. The United States priorities for maritime transportation in the Arctic region are:

a. To facilitate safe, secure, and reliable navigation;

b. To protect maritime commerce; and

c. To protect the environment.

2. Safe, secure, and environmentally sound maritime commerce in the Arctic region depends on infrastructure to support shipping activity, search and rescue capabilities, short- and long-range aids to navigation, high-risk area vessel-traffic management, iceberg warnings and other sea ice information, effective shipping standards, and measures to protect the marine environment. In addition, effective search and rescue in the Arctic will require local, State, Federal, tribal, commercial, volunteer, scientific, and multinational cooperation.

3. Working through the International Maritime Organization (IMO), the United States promotes strengthening existing measures and, as necessary, developing new measures to

improve the safety and security of maritime transportation, as well as to protect the marine environment in the Arctic region. These measures may include ship routing and reporting systems, such as traffic separation and vessel traffic management schemes in Arctic chokepoints; updating and strengthening of the Guidelines for Ships Operating in Arctic Ice-Covered Waters; underwater noise standards for commercial shipping; a review of shipping insurance issues; oil and other hazardous material pollution response agreements; and environmental standards.

4. Implementation: In carrying out this policy as it relates to maritime transportation in the Arctic region, the Secretaries of State, Defense, Transportation, Commerce, and Homeland Security, in coordination with heads of other relevant executive departments and agencies, shall:

a. Develop additional measures, in cooperation with other nations, to address issues that are likely to arise from expected increases in shipping into, out of, and through the Arctic region;

b. Commensurate with the level of human activity in the region, establish a risk-based capability to address hazards in the Arctic environment. Such efforts shall advance work on pollution prevention and response standards; determine basing and logistics support requirements, including necessary airlift and icebreaking capabilities; and improve plans and cooperative agreements for search and rescue;

c. Develop Arctic waterways management regimes in accordance with accepted international standards, including vessel traffic-monitoring and routing; safe navigation standards; accurate and standardized charts; and accurate and timely environmental and navigational information; and

d. Evaluate the feasibility of using access through the Arctic for strategic sealift and humanitarian aid and disaster relief.

G. Economic Issues, Including Energy

1. Sustainable development in the Arctic region poses particular challenges. Stakeholder input will inform key decisions as the United States seeks to promote economic and energy security. Climate change and other factors are significantly affecting the lives of Arctic inhabitants, particularly indigenous communities. The United States affirms the importance to Arctic communities of adapting to climate change, given their particular vulnerabilities.

2. Energy development in the Arctic region will play an important role in meeting growing global energy demand as the area is thought to contain a substantial portion of the world's undiscovered energy resources. The United States seeks to ensure that energy development throughout the Arctic occurs in an environmentally sound manner, taking into account the interests of indigenous and local communities, as well as open and transparent market principles. The United States seeks to balance access to, and development of, energy and other natural resources with the protection of the Arctic environment by ensuring that continental shelf resources are managed in a responsible manner and by continuing to work closely with other Arctic nations.

3. The United States recognizes the value and effectiveness of existing fora, such as the Arctic Council, the International Regulators Forum, and the International Standards Organization.

4. Implementation: In carrying out this policy as it relates to economic issues, including energy, the Secretaries of State, the Interior, Commerce, and Energy, in coordination with heads of other relevant executive departments and agencies, shall:

a. Seek to increase efforts, including those in the Arctic Council, to study changing climate conditions, with a view to preserving and enhancing economic opportunity in the Arctic region. Such efforts shall include inventories and assessments of villages, indigenous communities, subsistence opportunities, public facilities, infrastructure, oil and gas development projects, alternative energy development opportunities, forestry, cultural and other sites, living marine resources, and other elements of the Arctic's socioeconomic composition;

b. Work with other Arctic nations to ensure that hydrocarbon and other development in the Arctic region is carried out in accordance with accepted best practices and internationally recognized standards and the 2006 Group of Eight (G-8) Global Energy Security Principles;

c. Consult with other Arctic nations to discuss issues related to exploration, production, environmental and socioeconomic impacts, including drilling conduct, facility sharing, the sharing of environmental data, impact assessments, compatible monitoring programs, and reservoir management in areas with potentially shared resources;

d. Protect United States interests with respect to hydrocarbon reservoirs that may overlap boundaries to mitigate adverse environmental and economic consequences related to their development;

e. Identify opportunities for international cooperation on methane hydrate issues, North Slope hydrology, and other matters;

f. Explore whether there is a need for additional fora for informing decisions on hydrocarbon leasing, exploration, development, production, and transportation, as well as shared support activities, including infrastructure projects; and

g. Continue to emphasize cooperative mechanisms with nations operating in the region to address shared concerns, recognizing that most known Arctic oil and gas resources are located outside of United States jurisdiction.

H. Environmental Protection and Conservation of Natural Resources

1. The Arctic environment is unique and changing. Increased human activity is expected to bring additional stressors to the Arctic environment, with potentially serious consequences for Arctic communities and ecosystems.

2. Despite a growing body of research, the Arctic environment remains poorly understood. Sea ice and glaciers are in retreat. Permafrost is thawing and coasts are eroding. Pollutants from within and outside the Arctic are contaminating the region. Basic data are lacking in many fields.

High levels of uncertainty remain concerning the effects of climate change and increased human activity in the Arctic. Given the need for decisions to be based on sound scientific and socioeconomic information, Arctic environmental research, monitoring, and vulnerability assessments are top priorities.

For example, an understanding of the probable consequences of global climate variability and change on Arctic ecosystems is essential to guide the effective long-term management of Arctic natural resources and to address socioeconomic impacts of changing patterns in the use of natural resources.

3. Taking into account the limitations in existing data, United States efforts to protect the Arctic environment and to conserve its natural resources must be risk-based and proceed on the basis of the best available information.

4. The United States supports the application in the Arctic region of the general principles of international fisheries management outlined in the 1995 Agreement for the Implementation of the Provisions of the United Nations Convention on the Law of the Sea of December 10, 1982, relating to the Conservation and Management of Straddling Fish Stocks and Highly Migratory Fish Stocks and similar instruments.

The United States endorses the protection of vulnerable marine ecosystems in the Arctic from destructive fishing practices and seeks to ensure an adequate enforcement presence to safeguard Arctic living marine resources.

5. With temperature increases in the Arctic region, contaminants currently locked in the ice and soils will be released into the air, water, and land. This trend, along with increased human activity within and below the Arctic, will result in increased introduction of contaminants into the Arctic, including both persistent pollutants (e.g., persistent organic pollutants and mercury) and airborne pollutants (e.g., soot).

6. Implementation: In carrying out this policy as it relates to environmental protection and conservation of natural resources, the Secretaries of State, the Interior, Commerce, and Homeland Security and the Administrator of the Environmental Protection Agency, in coordination with heads of other relevant executive departments and agencies, shall:

a. In cooperation with other nations, respond effectively to increased pollutants and other environmental challenges;

b. Continue to identify ways to conserve, protect, and sustainably manage Arctic species and ensure adequate enforcement presence to safeguard living marine resources, taking account of the changing ranges or distribution of some species in the Arctic. For species whose range includes areas both within and beyond United States jurisdiction, the United States shall continue to collaborate with other governments to ensure effective conservation and management;

c. Seek to develop ways to address changing and expanding commercial fisheries in the Arctic, including through consideration of international agreements or organizations to govern future Arctic fisheries;

d. Pursue marine ecosystem-based management in the Arctic; and

e. Intensify efforts to develop scientific information on the adverse effects of pollutants on human health and the environment and work with other nations to reduce the introduction of key pollutants into the Arctic.

IV. Resources and Assets

A. Implementing a number of the policy elements directed above will require appropriate resources and assets. These elements shall be implemented consistent with applicable law and authorities of agencies, or heads of agencies, vested by law, and subject to the availability of appropriations. The heads of executive departments and agencies with responsibilities relating to the Arctic region shall work to identify future budget, administrative, personnel, or legislative proposal requirements to implement the elements of this directive. Policy on Protecting the Ocean Environment (PDD/NSC-36), Maritime Security Policy (NSPD-41/HSPD-13), and the National Strategy for Maritime Security (NSMS).[248]

End Notes

[1] Except for the subsection on the Arctic and the U.N. Convention on the Law of the Sea, this section was prepared by Ronald O'Rourke, Specialist in Naval Affairs, Foreign Affairs, Defense, and Trade Division.

[2] On November 25, 2008, voters in Greenland approved a referendum for greater autonomy that some observers view as a step toward eventual independence from Denmark. (Alan Cowell, "Greenland Vote Favors Independence," *New York Times*, November 26, 2008.)

[3] For more on the Arctic Council on the Internet, see http://www.arctic-council.org/.

[4] Title II of P.L. 98-373 is the National Critical Materials Act of 1984.

[5] As codified, the definition reads, "As used in this chapter.... "

[6] For discussions and (in some cases) maps, see Susan Joy Hassol, *Impacts of a Warming Arctic* [Executive Summary]. Cambridge, Cambridge University Press, 2004, p. 4. Available online at http://amap.no/acia/; Oran R. Yong and Niels Einarsson, *Arctic Human Development Report*, Stefansson Arctic Institute, Akureyri, Iceland, 2004, pp. 17-18, available online at http://www.svs.is/AHDR/AHDR%20chapters /English%20version /Chapters%20PDF.htm; and Hugo Ahlenius, editor in chief, et al. *Vital Arctic Graphics, People and Global Heritage on Our Last Wild Shores*, UNEP/GRID-Arendal, Arendal, Norway, p. 6, available online at http://www.grida.no/publications/vg/arctic/.

[7] A map showing the line that results from 10° isotherm definition is available online at https://www.cia.gov/library/publications/the-world-factbook/reference_maps/pdf/arctic. pdf.

[8] Discussion entitled "Geographical Coverage," available online at http://www.amap.no/ (click on "About AMAP" and then the tab "Geographical coverage.")

[9] Discussion entitled "Geographical Coverage," available online at http://www.amap.no/ (click on "About AMAP" and then the tab "Geographical coverage.")

[10] Title II of P.L. 98-373 is the National Critical Materials Act of 1984.

[11] These words are taken from the official title of P.L. 98-373. (Arctic Research and Policy Act of 1984 is the short title of Title I of P.L. 98-373.) The remainder of P.L. 98-373's official title relates to Title II of the act, the short title of which is the National Critical Materials Act of 1984.)

[12] The IARPC currently includes more than a dozen federal agencies, departments, and offices. Additional information on the IARPC is available on the Internet at http://www.nsf.gov/od/opp/arctic/iarpc/start.jsp.

[13] National Science Foundation, *FY 2013 Budget Request to Congress*, February 13, 2012, p. OPP-1 (pdf page 179 of 530). The document is available online at http://www nsf.gov/about/budget/fy2013/pdf/fy2013_rollup.pdf.

[14] National Science Foundation, *FY 2013 Budget Request to Congress*, February 13, 2012, p. OPP-1 (pdf page 179 of 530).

[15] National Science Foundation, *FY 2013Budget Request to Congress*, February 13, 2012, pp. OPP-2 and OPP-7 (pdf pages 180 and 185 of 530).

[16] CRS communication with State Department official, October 8, 2010.

[17] *National Security Strategy*, Washington, May 2010, p. 50. The quoted sentence constitutes the entirety of the document's comments specifically on the Arctic. It is the final sentence of a section on "sustain[ing] broad cooperation on key global challenges" that includes longer discussions on climate change, peacekeeping and armed conflict, pandemics and infectious disease, transnational criminal threats and threats to governance, and safeguarding the global commons.

[18] This section prepared by Marjorie Ann Browne, Specialist in International Relations, Foreign Affairs, Defense, and Trade Division.

[19] The United States is party to the four conventions adopted in 1958: Convention on the Territorial Sea and the Contiguous Zone, Convention on the High Seas, Convention on the Continental Shelf, and Convention on Fishing and Conservation of the Living Resources of the High Seas.

[20] Treaty Document 103-39.

[21] Other relevant provisions of the Convention, applicable depending on the extent of Arctic melting, relate to navigation, high seas freedoms, fisheries, and exclusive economic zones.

[22] The continental shelf is the under-sea extension of a coastal state's land territory. Article 76 of the Convention defines the continental shelf, *inter alia*, as "the seabed and subsoil of the submarine areas that extend beyond its [coastal state's] territorial sea throughout the natural prolongation of its land territory to the outer edge of the continental margin."

[23] A coastal State party has 10 years from the entry into force of the Convention for submission of information on its proposed limits. In May 2001, the Meeting of States Parties to the Convention decided that for any State for

which the Convention entered into force before May 13, 1999, the date of commencement of the 10-year time period for making submissions to the Commission is May 13, 1999.

[24] Annex II, Article 9. Article 83 of the Convention provides that questions relating to these boundary delimitation disputes shall be resolved by agreement between the States or by the Dispute Settlement options set forth in Part XV of the Convention.

[25] Article 76, para. 8.

[26] For more information, see http://www.state.gov/g/oes/continentalshelf/index.htm.

[27] This section prepared by Jane Leggett, Specialist in Energy and Environmental Policy, Resources, Science, and Industry Division.

[28] In scientific analyses, "ice-free" does not necessarily mean "no ice." The definition of "ice-free" or sea ice "extent" or "area" varies across studies. Sea ice "extent" is one common measure, equal to the sum of the area of grid cells that have ice concentration of less than a set percentage—frequently 15%. For more information, see the National Snow and Ice Data Center, http://nsidc.org/seaice/data/terminology.html.

[29] Muyin Wang and James E. Overland, "A Sea Ice Free Summer Arctic within 30 Years?," *Geophysical Research Letters* 36, no. L07502 (April 3, 2009): 10.1029/2009GL037820; Marika Holland, Cecilia M. Bitz, and Bruno Tremblay, "Future abrupt reductions in the summer Arctic sea ice," *Geophysical Research Letters* 33, no. L23503 (2006); But see also Julien Boé, Alex Hall, and Xin Qu, "Sources of spread in simulations of Arctic sea ice loss over the twenty-first century," *Climatic Change* 99, no. 3 (April 1, 2010): 637-645; I. Eisenman and J. S. Wettlaufer, "Nonlinear threshold behavior during the loss of Arctic sea ice," *Proceedings of the National Academy of Sciences* 106, no. 1 (January 6, 2009): 28-32; Dirk Notz, "The Future of Ice Sheets and Sea Ice: Between Reversible Retreat and Unstoppable Loss," *Proceedings of the National Academy of Sciences* 106, no. 49 (December 8, 2009): 20590-20595.

[30] Overland et al. state that "a warm Arctic-cold continent pattern represents a paradox of recent global warming: there is not a uniform pattern of temperature increases" due to a set of newly recognized processes described in Overland, J. E, K. R Wood, and M. Wang. "Warm Arctic-cold Continents: Climate Impacts of the Newly Open Arctic Sea." *Polar Research* 30 (2011). The authors raise a critical, unanswered question, "Is the observed severe mid-latitude weather in two adjacent years simply due to an extreme in chaotic processes alone, or do they included a partial but important Arctic forcing and connection due to recent changing conditions?" In other words, are recent patterns random anomalies, or might we expect more of the same?; among other examples, see also Lim, Young-Kwon, and Siegfried D. Schubert. "The Impact of ENSO and the Arctic Oscillation on Winter Temperature Extremes in the Southeast United States." *Geophysical Research Letters* 38, no. 15 (August 11, 2011): L15706.

[31] There was a regionally warm period in the Arctic from the mid-1920s to around 1940, which scientists have assessed to have been driven by natural climate variability. They have found that period to be distinctly different from the recent multi-decadal warming, in part because the early 20th century warming was concentrated in the northern high latitudes. See, for example, Figure 2, upper left graphic, in Geophysical Fluid Dynamics Laboratory, "Simulatoin of Early 20th Century Warming" at http://www.gfdl.noaa.gov/early-20th-century-global-warming.

[32] Steele, Michael, Wendy Ermold, and Jinlun Zhang. "Arctic Ocean Surface Warming Trends over the Past 100 Years." *Geophysical Research Letters* 35, no. 2 (January 29, 2008): L02614.

[33] J. Overland, J. Walsh, and M. Wang, *Arctic Report Card - Atmosphere* (NOAA Arctic Research Program, October 6, 2008), http://www.arctic.noaa.gov /reportcard /atmosphere. html.

[34] National Snow and Ice Data Center, "Arctic sea ice extent breaks 2007 record low" (August 27, 2012); Japanese Aerospace Exploration Agency, "A new record minimum of the Arctic sea ice extentn was set on 24 August 2012"; Arctic ROOS (Norway), "Daily Updated Time series of Arctic sea ice area and extent derived from SSMI data provided by NERSC" at http://arctic-roos.org/observations/satellite-data/sea-ice/ice-area-and-extent-in-arctic.

[35] William L. Chapman and John E. Walsh, "Simulations of Arctic Temperature and Pressure by Global Coupled Models," *Journal of Climate* 20, no. 4 (February 1, 2007): 609-632.

[36] See footnote 28. Also, although one Canadian scientist has predicted that recurrent ice-free summers may begin sometime between 2013 and 2020, this is not consistent with other climate models' projections.

[37] This section prepared by Carl Ek, Specialist in International Relations, Foreign Affairs, Defense, and Trade Division.

[38] "Russia, Canada Make Competing Claims To Arctic Resources," *The Canadian Press*, September 16, 2010.

[39] "5 Countries Agree To Talk, Not Compete, Over the Arctic," *New York Times*, May 29, 2008.

[40] For additional information, see Natalie Mychajlyszyn, *The Arctic: Geopolitical Issues*, Canadian Library of Parliament, October 24, 2008.

[41] "Russia, Norway Sign Deal On Barents Sea Border, Seek More Development in Mineral-rich Arctic," *Associated Press*, September 15, 2010.

[42] Cannon quoted in "Canada Seeks To Settle Arctic Borders," *Agence France Presse*, August 20, 2010. For additional information concerning Canada's August statement on Arctic policy, see "Statement on Canada's Arctic foreign policy: Exercising sovereignty and promoting Canada's Northern Strategy abroad," Foreign Affairs and International Trade Canada website, http://www.international.gc.ca/polar-polaire/canada_arctic_foreign_policyla_politique_etrangere_du_canada_pour_arctique.aspx?lang=eng.

[43] "Canada and Denmark Sign Arctic Cooperation Arrangement," *Targeted News Service*, May 17, 2010.

[44] This section prepared by John Frittelli, Specialist in Transportation Policy, Resources, Science, and Industry Division.

[45] Extended daylight hours in the Arctic during the summer may also be an advantage.

[46] Presentation by Stephen Carmel, Senior Vice President, Maersk Line Ltd., Halifax International Security Forum, *Arctic Security: The New Great Game?* November 21, 2009, available at http://fora.tv/.

[47] A third but more remote possibility is a route directly over the North Pole.

[48] In the summer of 2009, two German-owned ships are believed to be the first non-Russian ships to transit the NSR, sailing from Korea to deliver oil-service equipment to the Russian Arctic and then sailing on to Rotterdam.

[49] This was the route pioneered by the *SS Manhattan*, an oil tanker modified for ice breaking in 1969 to carry Alaskan North Slope oil to the Atlantic. This was the first commercial passage through the NWP, but the building of the Alaskan pipeline was found to be the more economical means of transporting oil from the North Slope to the lower 48 states.

[50] Although the NWP is often compared to the alternative route through the Panama Canal in terms of distance and sailing days from Asia to the U.S. east coast, another alternative to consider is the shorter and faster transcontinental rail route across Canada or the United States. The Panama Canal is undergoing an expansion project, expected to be completed by 2015, to allow larger ships with about three times the cargo capacity to pass through.

[51] ITAR-TASS, "Rosatomflot Ends Shipping Season Along Northern Sea Route," November 29, 2011.

[52] "U.S. Needs to Prepare for Arctic Traffic Surge," *Anchorage Daily News*, February 12, 2008, p. A1.

[53] "Northern Waters," *The Globe and Mail* (Canada), August 14, 2008, p. A8.

[54] See http://www.aeco.no/index.htm for more information.

[55] S.E.L. Howell and J.J. Yackel, "A Vessel Transit Assessment of Sea Ice Variability in the Western Arctic, 1969-2002: Implications for Ship Navigation," *Canadian Journal of Remote Sensing*, vol. 30, no. 2, 2004.

[56] "Arctic Unlikely to See Major Shipping Growth," *New Zealand Transport and Logistics Business Week*, April 24, 2008.

[57] Presentation by Stephen Carmel, Senior Vice President, Maersk Line Ltd., Halifax International Security Forum, *Arctic Security: The New Great Game?* November 21, 2009, available at http://fora.tv/.

[58] In July and August 2010, NOAA surveyed the Bering Straits area in order to update its charts but stated that it will take more than 25 years to map the prioritized areas of navigational significance in U.S. Arctic waters. See http://www.noaanews.noaa.gov/stories 2010/20100720_fairweather.html.

[59] Ice reporting that currently exists is intended for scientists not mariners.

[60] For further information, see http://www.poa.usace.army.mil/en/cw/AKPortsStudy.htm, and FY2013 USACE Budget Justification, p. POD-5.

[61] See http://www.cmts.gov/Activities/ActionTeams.aspx.

[62] See http://www.imo.org/ for more information.

[63] See MSC/Circ. 1056/MEPC/Circ.399 at http://www.imo.org/.

[64] Edwin H. Anderson, "Polar Shipping, The Forthcoming Polar Code and Implications for the Polar Environments," *Journal of Maritime Law and Commerce,* v. 43, no. 1, January 2012.

[65] The Coast Guard is studying shipping routes through the Bering Strait for possible safety enhancements. See 75 FR 68568, November 8, 2010.

[66] This section prepared by Curry Hagerty, Specialist in Energy and Natural Resources Policy, Resources, Science, and Industry Division; Peter Folger, Specialist in Energy and Natural Resources Policy, Resources, Science, and Industry Division; and Marc Humphries, Analyst in Energy Policy, Resources, Science, and Industry Division.

[67] The Continuing Appropriations Resolution 2009 (P.L. 110-329) did not extend the annual congressional moratorium on oil and gas leasing activities. On March 11, 2009, the Omnibus Appropriations Act, 2009 (P.L. 111-8) was enacted without moratoria provisions, confirming that the oil and gas development moratoria in the OCS along the Atlantic and Pacific Coasts, parts of Alaska, and the Gulf of Mexico that had been in place since 1982 had not been restored in 2009 appropriations measures. The combined effect of Presidential Directives from 1990 to 2008 also impacted moratoria constraints.

[68] On July 14, 2008, a Modification of the Presidential Withdrawal of areas of the United States Outer Continental Shelf from leasing disposition was announced by President Bush in the following statement, "Under the authority vested in me as President of the United States, including section 12(a) of the Outer Continental Shelf Lands Act, 43 U.S.C. 1341(a), I hereby modify the prior memoranda of withdrawals from disposition by leasing of the United States Outer Continental Shelf issued on August 4, 1992."

[69] Effective on June 29, 2009, the rule established two types of leases: (1) commercial leases for full development and power generation, and (2) limited leases for resource assessment and technology testing. See 73 Fed Reg 39376.

[70] A seismic array is typically a long string or streamer of geophones—acoustic devices used for recording seismic signals—towed behind a ship while the ship traverses a prospective oil and gas-bearing portion of the seafloor. The seismic signals are processed and interpreted to give a cross-section or three-dimensional image of the subsurface.

[71] See USGS Circum-Arctic Resource Appraisal website at http://energy.usgs.gov/arctic/.

[72] USGS Fact Sheet 2008-3049: Circum-Arctic Resource Appraisal: Estimates of Undiscovered Oil and Gas North of the Arctic Circle; at http://pubs.usgs.gov/fs/2008/3049/.

[73] USGS Fact Sheet 2008-3049: Circum-Arctic Resource Appraisal: Estimates of Undiscovered Oil and Gas North of the Arctic Circle; at http://pubs.usgs.gov/fs/2008/3049/.

[74] Hugo Miller, "BHP Billiton Leads Arctic Gold Hunt in Global Warming Bonanza," Bloomberg.com (April 16, 2008), at http://www.bloomberg.com/apps/news?pid= 20601081&sid=ag.kQZln.mFg&refer=australia.

[75] The Associated Press, "Greenland Opens to Oil Firms; Melting Ice Unlocks Reserves," International Herald Tribune (January 14, 2008), at http://www.iht.com/articles/ap /2008/01/14/business/NA-FIN-US-Greenland-Oil-Companies.php.

[76] Tony Halpin, "President Medvedev Threatens Russian Arctic Annexation," Times Online (September 18, 2008), at http://www.timesonline.co.uk/tol/news/world/europe/article 4773567.ece.

[77] The purpose of the U.S. Extended Continental Shelf Project is to establish the full extent of the continental shelf of the United States, consistent with international law. Involved in this mission are the U.S. Coast Guard (USCG), Department of State (DOS), and the University of New Hampshire (UNH). NOAA has the lead in collecting bathymetric data. The U.S. Geological Survey has the lead in collecting seismic data.

[78] See USGS, U.S. Extended Continental Shelf Project, at http://www.state.gov/g/oes /continentalshelf/.

[79] Press release, Office of the Prime Minister, Canada (August 26, 2008), at http://www.pm.gc.ca/eng/media.asp?category=1&id=2242.

[80] National Security Presidential Directive/NSPD-66 and Homeland Security Presidential Directive/HSPD-25, at http://www.whitehouse.gov/news/releases/2009/01/20090112-3.html.

[81] Don Gautier, Research Geologist, USGS, quoted in "Countries in Tug-of-War Over Arctic Resources," CNN.com (January 2, 2009), at http://www.cnn.com/2009/TECH/science /01/02/arctic.rights.dispute/index.html?iref=newssearch.

[82] Carolyn Fitzpatrick, "Heavy Haul in the High North," Railway Gazette International (July 24, 2008), at http://www.railwaygazette.com/news/single-view/view/10/heavy-haul-in-the-high-north.html.

[83] See National Oceanic and Atmospheric Administration, Arctic Change, at http://www.arctic.noaa.gov/detect /landroad.shtml?page=land.

[84] This section prepared by Jonathan L. Ramseur, Specialist in Environmental Policy, Resources, Science, and Industry Division.

[85] Arctic Monitoring and Assessment Programme (AMAP), Arctic Oil and Gas 2007 (2008).

[86] Arctic Monitoring and Assessment Programme (AMAP), Arctic Oil and Gas 2007 (2008).

[87] See this report's section "Implications for Sea Transportation," by John Fritelli.

[88] For example, the *Exxon Valdez* spilled approximately 11 million gallons of oil, but its carrying capacity was approximately 60 million gallons.

[89] See this report's section "Implication of Changes in the Arctic for Oil, Gas, and Mineral Exploration and Development," by Peter Folger and Marc Humphries.

[90] Larger oil spills occurred during the 1991 Iraq War, but many of those spills were deliberate. A 1910-1911 onshore oil blowout in the California San Joaquin Valley is reported to have spilled 9.4 million barrels of crude oil (almost 400 million gallons).

[91] See CRS Report R41531, *Deepwater Horizon Oil Spill: The Fate of the Oil*, by Jonathan L. Ramseur.

[92] National Research Council (NRC) of the National Academies of Science, *Oil in the Sea III: Inputs, Fates, and Effects* (2003).

[93] See CRS Report RL33705, *Oil Spills in U.S. Coastal Waters: Background and Governance*, by Jonathan L. Ramseur; and Dagmar Etkin (Environmental Research Consulting), Analysis of U.S. Oil Spillage, Prepared for American Petroleum Institute, August 2009.

[94] National Research Council of the National Academies of Science, *Cumulative Environmental Effects of Oil and Gas Activities on Alaska's North Slope* (2003).

[95] See, for example, Minerals Management Service (MMS), Outer Continental Shelf Oil & Gas Leasing Program: 2007-2012, Final Environmental Impact Statement, April 2007, Chapter 4; MMS, Proposed Gulf of Mexico OCS Oil and Gas Lease Sale 206, Central Planning Area, Environmental Assessment, October 2007;

[96] Letter from Marvin E. Odum, President, Shell Oil Company to S. Elizabeth Birnbaum, Minerals Management Service (May 14, 2010). Cited in a staff paper from the National Commission on the BP Deepwater Horizon Oil Spill and Offshore Drilling ("The Challenges of Oil Spill Response in the Arctic," January 2011).

[97] AMAP, *Arctic Oil and Gas 2007 (2008).*

[98] AMAP, *Arctic Oil and Gas 2007* (2008).

[99] World Wildlife Fund, *Oil Spill: Response Challenges in Arctic Waters* (2007).

[100] Coastal Response Research Center, *Opening the Arctic Seas: Envisioning Disasters and Framing Solutions* (2009), partnership between the National Oceanic and Atmospheric Administration and the University of New Hampshire.

[101] Although the response gap in the Arctic has not been quantified, a recent estimate of Prince William Sound (PWS) may be instructive. A 2007 study found a response gap for PWS of 38% for the time of the study period (65% during the winter season). Note that PWS has existing infrastructure for response, while the more remote Arctic areas do not. Nuka Research and Planning Group, LLC, *Response Gap Estimate for Two Operating Areas in Prince William Sound, Alaska* (2007), Report to Prince William Sound Regional Citizens' Advisory Council.

[102] Arthur Brooks (Commander, 17th Coast Guard District) Coast Guard Journal, Arctic Journal (April 7, 2008).

[103] U.S. Coast Guard, *Report to Congress: U.S. Coast Guard Polar Operations*, December 2008.

[104] A single road connects Deadhorse, Alaska and the Prudhoe Bay with central Alaska (Fairbanks). For more details and maps of the area, see Nuka Research and Planning Group, *Oil Spill Prevention and Response in the U.S. Arctic Ocean: Unexamined Risks, Unacceptable Consequences*, Commissioned by Pew Environment Group, November 2010.

[105] Government Accountability Office, *Coast Guard: Efforts to Identify Arctic Requirements Are Ongoing, but More Communication about Agency Planning Efforts Would Be Beneficial*, GAO-10-870, September 2010.

[106] Transportation Research Board of the National Academy of Sciences, *Risk of Vessel Accidents and Spills in the Aleutian Islands: Designing a Comprehensive Risk Assessment* (2008), Special Report 293, National Academies Press. Washington, DC.

[107] Coastal Response Research Center, *Opening the Arctic Seas: Envisioning Disasters and Framing Solutions*, (2009), partnership between the National Oceanic and Atmospheric Administration and the University of New Hampshire. See also, U.S. Arctic Research Commission, (Draft) *White Paper: U.S. Arctic Research Commission Recommends Steps to Expanded U.S. Funding for Arctic/Subarctic Oil Spill Research* (February 24, 2010).

[108] Arctic Monitoring and Assessment Programme (AMAP), *Arctic Oil and Gas 2007* (2008).

[109] National Research Council, *Cumulative Environmental Effects of Oil and Gas Activities on Alaska's North Slope* (2003).

[110] Arturo A. Keller and Kristin Clark, *Oil Recovery with Novel Skimmer Surfaces under Cold Climate Conditions* (2007), prepared for the Minerals Management Service.

[111] World Wildlife Fund, *Oil Spill: Response Challenges in Arctic Waters* (2007). For further discussion of issues relating to oil spills, see CRS Report RL33705, *Oil Spills in U.S. Coastal Waters: Background and Governance*, by Jonathan L. Ramseur.

[112] Coastal Response Research Center, *Opening the Arctic Seas: Envisioning Disasters and Framing Solutions*, (2009), partnership between the National Oceanic and Atmospheric Administration and the University of New Hampshire.

[113] This section prepared by Eugene Buck, Specialist in Natural Resources Policy, Resources, Science, and Industry Division.

[114] Erik J. Molenaar and Robert Corell, *Arctic Fisheries*, Arctic Transform, February 9, 2009; available at http://arctictransform.org/download/FishBP.pdf.

[115] 74 *Federal Register* 56734-56746, November 3, 2009.

[116] Information on the symposium is available at http://www.nprb.org/iafs2009/.

[117] The website is available at http://alaskafisheries.noaa.gov/sustainablefisheries/arctic/.

[118] Prepared by Lynne Corn and Eugene Buck, Specialists in Natural Resources Policy, Resources, Science, and Industry Division.

[119] Fish and Wildlife Service, "Endangered and Threatened Wildlife and Plants; Review of Native Species That Are Candidates for Listing as Endangered or Threatened; Annual Notice of Findings on Resubmitted Petitions; Annual Description of Progress on Listing Actions; Proposed Rule," 76 *Federal Register* 66370-66439, October 26, 2001.

[120] 73 *Federal Register* 79822-79828.

[121] 75 *Federal Register* 65239-65248.

[122] 74 *Federal Register* 53683-58696, October 20, 2009.

[123] 75 *Federal Register* 77476-77495.

[124] 75 *Federal Register* 77496-77515.

[125] Among biologists, it is traditionally said that a species faced with extreme change can respond in three basic ways: "migrate, mutate, or die." When change is rapid enough, mutation (accompanied by natural selection of individuals within the population more suited to the changed environment) may not be able to occur fast enough, leaving migration and death as the only options. The problem of response rate is more severe for species that reproduce slowly (e.g., polar bears) and less severe for species that reproduce rapidly (e.g., algae).

[126] The efficacy and the effect of this tactic is often questioned, since natural migration is unlikely to involve the entire suite of species in an ecosystem (e.g., host plants might not move north (or up) as fast as their moth herbivores, nor as fast as the birds that depend on the moths). Moreover, the southerners will not find a land of sterile bare dirt—the species that are already there may be threatened themselves by the competition from the new arrivals, perhaps tipping the balance and pushing still more species toward extinction.

[127] Under the MMPA, both NMFS and FWS have responsibility for additional marine mammal species (e.g., manatees, sea otters, dolphins) which are not currently found in the Arctic.

[128] This section was originally prepared by Roger Walke, who was a Specialist in American Indian Policy, Domestic Social Policy Division, until his retirement from CRS in October 2010. It has been updated by Jane A. Leggett, Specialist in Environmental and Energy Policy in CRS's Resources, Science and Industry Division.

[129] *Arctic Human Development Report*, ed. Joan Nymand Larsen, et al. (Akureyri, Iceland: Stefansson Arctic Institute, 2004), p. 47; this report is subsequently cited in this section as *AHDR*. The seven countries are Canada, Denmark-Greenland, Finland, Norway, Russia, Sweden, and the United States.

[130] John F. Hoffecker, *A Prehistory of the North: Human Settlement of the Higher Latitudes* (New Brunswick, NJ: Rutgers University Press, 2005), pp. 8, 81, 112-115.

[131] *AHDR*, pp. 47, 53; David Crystal, *Cambridge Encyclopedia of Language*, 2nd ed. (Cambridge: Cambridge University Press, 1997), chap. 50; *Ethnologue: Languages of the World*, 16th ed., ed. M. Paul Lewis (Dallas: SIL International, 2009) available online at http://www.ethnologue.com/. The number of languages and language families varies not only with definitions of the Arctic but with definitions of languages and language families.

[132] Jim Berner, et al., *Arctic Climate Impact Assessment* (Cambridge: Cambridge University Press, 2005), chapter 12; this report is subsequently cited in this section as *ACIA*.

[133] *ACIA*, pp. 654-655.

[134] Fennoscandia refers to the Scandinavian Peninsula, Finland, the Kola Peninsula of Russia, and certain parts of Russia bordering on Finland.

[135] Janet Martin, *Treasure in the Land of Darkness: The Fur Trade and Its Significance for Medieval Russia* (Cambridge: Cambridge University Press, 1986), pp. 41-42; James Forsyth, *A History of the Peoples of Siberia: Russia's North Asian Colony, 1581-1990* (Cambridge: Cambridge University Press, 1992), pp. 69-83, 102; Lassi K. Heininen, "Different Images of the Arctic and the Circumpolar North in World Politics," in *Knowledge and Power in the Arctic*, Proceedings at a Conference in Rovaniemi, April 16-18, 2007, Arctic Centre Reports 48, ed. Paula Kankaanpaa, et al. (Rovaniemi, Finland: University of Lapland, Arctic Centre, 2007), p. 125.

[136] James W. VanStone, "Exploration and Contact History of Western Alaska," and David Damas, "Copper Eskimo," and J. Garth Taylor, "Historical Ethnography of the Labrador Coast," in *Handbook of North American Indian: Vol. 5, Arctic,* vol. ed. David Damas, gen. ed. William C. Sturtevant (Washington: Smithsonian, 1984), pp. 149-155, 408, 509- 510.

[137] Inge Kleivan, "History of Norse Greenland," in *Handbook, Vol. 5, Arctic, op. cit.,*, pp. 549-555; Finn Gad, "Danish Greenland Policies," in *Handbook of North American Indians: Vol. 4, History of Indian-White Relations*, vol. ed. Wilcomb E. Washburn, gen. ed. William C. Sturtevant (Washington: Smithsonian, 1988), p. 110.

[138] Parkinson, Alan J. The Arctic Human Health Initiative. Washington DC: U.S. Department of Health and Human Services, 2006.

[139] *AHDR*, pp. 19, 29. Estimates of Arctic indigenous populations are complicated by varying definitions not only of the Arctic but also of indigenous peoples; for instance, Russia does not count some non-European Arctic ethnic groups, such as the Yakut, as "indigenous minorities" (see "Peoples of the Arctic: Characteristics of Human Populations Relevant to Pollution Issues," in *AMAP Assessment Report: Arctic Pollution Issues*, ed. Simon J. Wilson et al. (Oslo: Arctic Monitoring and Assessment Programme, 1998), pp. 167-169; this report is subsequently cited in this section as *AMAP 1998*.

[140] *AMAP 1998*, chapter 5; see also Birger Poppel et al., *SLiCA Results*, Survey of Living Conditions in the Arctic (Anchorage: Institute of Social and Economic Research, University of Alaska Anchorage, 2007), pp. 4-7, http://www.arcticlivingconditions.org.

[141] Annika E. Nilson and Henry P. Huntington, *Arctic Pollution 2009* (Oslo: Arctic Monitoring and Assessment Programme, 2009), p. 39-41; this report is subsequently cited in this section as *AMAP 2009*.

[142] *ACIA*, p. 1000.

[143] *SLiCA Results, op.cit.*, pp. v, 4-8.

[144] *AHDR*, p. 232.

[145] *AHDR*, chapter 4, and pp. 232-233.

[146] *AHDR*, chapters 6-7, and pp. 232-233.

[147] *ACIA*, pp. 1000-1001, 1004.

[148] *ACIA*, pp. 1000-1001, 1004.

[149] Parkinson, Alan J. *The Arctic Human Health Initiative*. Washington DC: U.S. Department of Health and Human Services, 2006; Brubaker, Michael, James Berner, Raj Chavan, and John Warren. "Climate Change and Health Effects in Northwest Alaska." *Global Health Action* 4 (October 18, 2011).

[150] U.S. Government Accountability Office (GAO), *Alaska Native Villages: Villages Affected by Flooding and Erosion Have Difficulty Qualifying for Federal Assistance*, GAO-04-895T, June 29, 2004, p. i, http://www.gao.gov/new.items/ d04895t.pdf. See also, Government Accountability Office, *Alaska Native Villages: Most Are Affected by Flooding and Erosion, but Few Qualify for Federal Assistance*, GAO-04-142, December 12, 2003, http://www.gao.gov/new.items/ d04142.pdf.

[151] U.S. Army Corps of Engineers. Alaska Baseline Erosion Assessment (BEA), March 2009.

[152] GAO, *Alaska Native Villages: Limited Progress Has Been Made on Relocating Villages Threatened by Flooding and Erosion*, June 3, 2009.

[153] Alaska Federation of Natives, Human Resources Committee, *2010 Federal Priorities* (Anchorage: Alaska Federation of Natives, 2010), pp. 22-23; available at http://www.nativefederation.org/documents/ 2010_AFN_Federal_Priorities.pdf. See also, Government Accountability Office, *Alaska Native Villages: Limited Progress Has Been Made on Relocating Villages Threatened by Flooding and Erosion*, GAO-09-551, June 3, 2009, http://www.gao.gov/new.items/d09551.pdf. For a more detailed anecdote, see a 2012 interview with Brice Eningowuk, Shishmaref Erosion and Relocation Committee member at Shearer, Christine. "Climate Crisis: Alaskan Village Shishmaref Sinking Into the Sea." *Imagined Magazine*, July 8, 2012. http://imaginedmag.com/2012/07/climate-crisisalaskan-village-shishmaref-sinking-into-the-sea/.

[154] Press Office of Senator Lisa Murkowski. "Murkowski Urges Greater Tribal Consultation from Administration; Senator Spotlights 'Monumental' Climate Struggles Faced by Alaska Natvies, Coastal Communities", July 19, 2012.

[155] *ACIA*, pp. 1001, 1004.

[156] Aqqaluk Lynge, Chair, Inuit Circumpolar Council, quoted in Stephanie McFeeters, "Lynge talks future of Inuit people," *The Dartmouth*, February 8, 2012.

[157] See, for example, "Health: Increased Bacterial Loads in Potable Water Could Have Significant Health Effects on Indigenous People From the Arctic to Uganda, Says Vanier Scholar." National Aboriginal Health Organization (NAHO), February 17, 2012. http://www.naho.ca/blog/2012/02/17/health-increased-bacterial-loads-in-potable-

watercould-have-significant-health-effects-on-indigenous-people-from-the-arctic-to-uganda-says-vanier-scholar/; or, Kallenborn, et al. *Combined Effects of Selected Pollutants and Climate Change in the Arctic Environment.* Oslo, Norway: Arctic Monitoring and Assessment Programme (AMAP), Arctic Council, 2011.

[158] *AMAP Assessment 2009: Human Health in the Arctic*, ed. Simon J. Wilson and Carolyn Symon (Oslo: Arctic Monitoring and Assessment Programme, 2009), pp. 4-6, 143.

[159] Brubaker, Michael, James Berner, Raj Chavan, and John Warren. "Climate Change and Health Effects in Northwest Alaska." *Global Health Action* 4 (October 18, 2011); John Warren, "Climate change could affect human health," *Mukluk Telegraph*, January/February 2005, pp. 5-6.

[160] John Warren, "Climate change could affect human health," *Mukluk Telegraph*, January/February 2005, pp. 5-6.

[161] *AHDR*, p. 235.

[162] See http://www.arctic-council.org/. The six organizations are the Aleut International Association, Arctic Athabaskan Council, Gwich'in Council International, Inuit Circumpolar Council, RAIPON (Russian Association of Indigenous Peoples of the North), and Saami Council.

[163] United Nations, Declaration on the Rights of Indigenous Peoples, Resolution A/RES/61/295, at http://social.un.org/index/IndigenousPeoples/DeclarationontheRightsofIndigenous Peoples.aspx.

[164] See http://www.indigenoussummit.com/servlet/content/home.html.

[165] K. Galloway-McLean et al., *Report of the Indigenous Peoples' Global Summit on Climate Change: 20-24 April 2009, Anchorage, Alaska* (Darwin, Australia: United Nations University—Traditional Knowledge Initiative, 2009), pp. 5-7; available at http://www.indigenoussummit.com/servlet/content/home.html.

[166] This section prepared by Ronald O'Rourke, Specialist in Naval Affairs, Foreign Affairs, Defense, and Trade Division. It is adapted from CRS Report RL34391, *Coast Guard Polar Icebreaker Modernization: Background and Issues for Congress*, by Ronald O'Rourke.

[167] The Coast Guard's polar icebreakers also support U.S. scientific research in the Antarctic.

[168] *United States Coast Guard High Latitude Region Mission Analysis Capstone Summary*, July 2010, pp. 12-13, 15.

[169] Source: Transcript of hearing.

[170] This section prepared by Ronald O'Rourke, Specialist in Naval Affairs, Foreign Affairs, Defense, and Trade Division.

[171] Dan Elliott, "NORAD Chiefs Eye Busier Arctic," *Fairbanks Daily News-Miner*, May 7, 2011.

[172] For a report assessing certain emergency scenarios in the Arctic, including search and rescue scenarios, see *Opening the Arctic Seas, Envisioning Disasters and Framing Solutions*, Coastal Response and Research Center, University of New Hampshire, report of January 2009, based on conference held March 18-20, 2008, at Durham, New Hampshire.

[173] Cid Standifer, "Coast Guard Comandant: Service Still Committed To Eight NSCs," Inside the Navy, January 17, 2011.

[174] Jennifer Scholtes, "Papp: Coast Guard Arctic Plan Must Look Beyond Icebreakers," *CQ Homeland Security*, March 1, 2011.

[175] Source: Transcript of hearing.

[176] Source: State Department website accessed at http://www.state.gov/g/oes /ocns/opa /arc /c29382.htm on July 7, 2011.

[177] The guidelines, as stated in Article 7, are as follows:

(a) search and rescue operations conducted pursuant to this Agreement in the territory of a Party shall be carried out consistent with the laws and regulations of that Party;

(b) if a search and rescue agency and/or RCC of a Party receives information that any person is, or appears to be, in distress, that Party shall take urgent steps to ensure that the necessary assistance is provided;

(c) any Party having reason to believe that a person, a vessel or other craft or aircraft is in a state of emergency in the area of another Party as set forth in paragraph 2 of the Annex shall forward as soon as possible all available information to the Party or Parties concerned;

(d) the search and rescue agency and/or RCC of a Party that has received information concerning a situation provided for in subparagraph (b) of this paragraph may request assistance from the other Parties;

(e) the Party to whom a request for assistance is submitted shall promptly decide on and inform the requesting Party whether or not it is in a position to render the assistance requested and shall promptly indicate the scope and the terms of the assistance that can be rendered;

(f) the Parties shall ensure that assistance be provided to any person in distress. They shall do so regardless of the nationality or status of such a person or the circumstances in which that person is found; and

(g) a Party shall promptly provide all relevant information regarding the search and rescue of any person to the consular or diplomatic authorities concerned.

[178] Article 20 identifies the government of Canada as the depository for the agreement.

[179] Source: Text of final version of agreement made ready for signing and dated April 21, 2011, accessed July 7, 2011, at http://arctic-council.org/filearchive/Arctic_SAR_Agreement_ EN_FINAL_for_signature_21-Apr-2011.pdf

[180] Source: State Department website accessed at http://www.state.gov/r/pa/prs /ps/2011/05 /163285.htm on July 7, 2011.

[181] This section prepared by Carl Ek, Specialist in International Relations, Foreign Affairs, Defense, and Trade Division.

[182] "Arctic Sea Ice Continues to Decline, Hits Near Record Low," *IB Times*, October 4, 2011.

[183] "China, India Enter Heating-up Arctic Race," *Asia Times,* January 25, 2012. "Aston in Far North to Reach EU Policy on Arctic," *Agence France Presse,* March 8, 2012. "EU Application to Arctic Council Postponed," North Norway European Office, May 18, 2011 http://www.northnorway.org/nyheter/847-eu-application-to-arctic-councilpostponed See also: "Arctic Governance: Balancing Challenges and Development," European Parliament Regional Briefing 2012, DG EXPO/B/PolDep/Note/2012_136 June, 2012

[184] Conference on the Arctic Ocean, May 26, 2008, Arctic Council website: http://arctic-council.org/article/2008/5/conference_on_the_arctic_ocean.

[185] Conference In Ilulissat, Greenland: Landmark Political Declaration on the Future of the Arctic, Edited December 11, 2008, website of the Danish Foreign Ministry, http://www.um.dk/en.

[186] The Council members are the "Arctic 5" coastal States plus Sweden, Finland, and Iceland. Six indigenous Arctic peoples organizations are permanent participants. Several other countries, including Spain, China, and the UK, have observer status. For additional information, see the Arctic Council's website: http://www.arctic-council.org/index.php/ en/. "Document On Cooperation In Arctic Search and Rescue Efforts Could Be Ready In April 2011," *Interfax: Russia & CIS Military Newswire*, February 26, 2010.

[187] "States Set Rules on Exploiting Arctic Wealth," *Agence France Presse*, May 12, 2011.

[188] "Secretary Clinton Signs the Arctic Search and Rescue Agreement with Other Arctic Nations," U.S. Department of State website, May 12, 2011. "Warming Arctic Opens Way to Competition For Resources," *Washington Post*, May 16, 2011.

[189] "Clinton in Arctic to See Impact of Climate Change," *Agence France Presse,* June 2, 2012.

[190] During an October 2011 conference on Arctic shipping, President Putin enthused that "the Arctic is the shortcut between the largest markets of Europe and the Asia-Pacific region.... It is an excellent opportunity to optimize costs." "Warming Revives Dream of Sea Route in Russian Arctic," *New York Times,* October 17, 2011.

[191] "Foreign Ministers of Russia and Canada Agreed To Go To the Ice Without the Third Parties," *WPS: Defense & Security,* September 20, 2010. "Russia Will Staunchly Defend its Interests in the Arctic Region – Putin," *Interfax*. June 30, 2011. "Russia to Deploy Troops to Defend Interests in Arctic," *Wall Street Journal*, July 2, 2011. "Russia, U.S., Norway to Hold Joint Naval Drills in August," *RIA Novosti,* May 24, 2012 http://en.rian.ru/world/20120524/ 173654802.html.

[192] "The Dragon Eyes the Top of the World," David Curtis Wright, Naval War College, China Maritime Studies Institute, Number 8, August 2011.

[193] "China Defines Itself as a 'Near-Arctic State,' Says SIPRI," Swedish International Peace Research Institute, May 10, 2012. "China's Arctic Ambitions," *Stratfor,* June 18, 2012. "China: Icebreaking in the Arctic," *The ISN Blog,* May 4, 2012. "Ice Station Dragon: China's Strategic Arctic Interest," *Defense News*, May 16, 2011.

[194] Heather Conley, "The Colder War: U.S., Russia and Others Are Vying for Control of Santa's Back Yard," *Washington Post,* December 23, 2011.

[195] "Welcome To a New Era of Arctic Security," Rob Huebert, Canadian Defense and Foreign Affairs Institute, August, 2010.

[196] See, for example, "Canada Vows 'Firm' Defence of Arctic Border," *Canwest News Service,* November 24, 2009. "Danes With Dogs To Join Military Sovereignty Patrol Of Canadian Arctic," *The Canadian Press,* March 3, 2010. "Norway Plans One New Arctic Base, Perhaps More," *Defense News,* March 8, 2010.

[197] "The Arctic: Special Report," *The Economist,* June 16, 2012. p. 10.

[198] "A New Security Architecture for the Arctic: An American Perspective," Center for Strategic and International Studies (CSIS), January 2012, p. 30.

[199] Nonetheless, one analyst has noted that the word "Arctic" does not appear either in the 2010 NATO Strategic Concept, nor in the Summit Declaration of the 2012 Chicago Summit – likely at the insistence of Canada. See "NATO in the Arctic: Challenges and Opportunities," by Luke Coffee, Heritage Foundation Issue Brief No. 3646, June 22, 2012.

[200] "Russia, Norway Sign Border Deal For Arctic Energy," *Reuters*, September 15, 2010. "NATO Chief Cautions Against Division Over Arctic," *Canwest News Service*, January 29, 2009. "NATO Proposes Arctic Cooperation With Russia," Reuters, January 29, 2009. Speech by NATO Secretary General Jaap de Hoop Scheffer On Security Prospects In the High North, January 29, 2009, http://www.nato.int/docu/speech/2009/s090129a.html. "NATO Chief Wars Of Climate Change Security Risks," *Agence France Presse*, October 1, 2009. "Rogozin Says He Won't Discuss Cooperation In Arctic With NATO," *Interfax: Russia & CIS General Newswire,* March 27, 2009.

[201] "NATO Tests Its Forces in Arctic," February 16, 2012, Atlantic Council web page: http://www.acus.org/natosource/nato-tests-its-forces-arctic "NATO, Russia Stage Arctic War Games," April 25, 2012 Atlantic Council web page: http://www. acus.org/print/68922 See also Cold Response 2012, Norwegian Armed Forces web page: http://mil.no/excercises/coldresponse2012/pages/default.aspx

[202] "The Arctic: Special Report," *The Economist,* June 16, 2012. p. 11.

[203] "Russia Will Protect Interests In Arctic: Official," *Agence France Presse,* June 10, 2009. "Russia Opposes Arms Race In Arctic Region—Diplomat," *ITAR-TASS World Service,* July 21, 2009.

[204] "China Prepares For An Ice-free Arctic," Linda Jakobson, *SIPRI Insights On Peace and Security*, No. 2010/2, March 2010. "Admiral Urges Government To Stake Claim In the Arctic," *South China Morning Post*, March 6, 2010. "Norway Wants to Block China from Arctic Council Over 2010 Nobel Peace Prize Row," *Agence France Presse*, January 25, 2012. "Clinton in Arctic to See Impact of Climate Change," *Agence France Presse,* June 2, 2012.

[205] "A New Security Architecture for the Arctic: An American Perspective," Center for Strategic and International Studies (CSIS), January 2012, p. 37.

[206] This section prepared by Ronald O'Rourke, Specialist in Naval Affairs, Foreign Affairs, Defense, and Trade Division.

[207] For a recent article concerning Soviet submarine operations in the Arctic during the Cold War, see Bob Weber, "Russian Maps Suggest Soviet Subs Cruised Canadian Arctic," *The Globe and Mail (www.theglobeandmail.com)*, December 6, 2011.

[208] See, for example, Associated Press, "Military Powers Beef Up Arctic Presence," *The Korea Times (www.koreatimes.co.kr)*, April 17, 2012, accessed June 15, 2012 at http://www.koreatimes.co.kr/www/news/tech/2012/ 06/129_109145.html; Alan Cullison, "Russia To Deploy Troops To Defend Interests In Arctic," *Wall Street Journal*, July 2, 2011: 2; David Pugliese, "[Canadian] Senators: Arm [Canadian] Coast Guard Patrols in Canada's Arctic," *Defense News*, January 18, 2010: 12.

[209] See, for example, John Vandiver, "NATO Commander Sees Arctic Seabed As Cooperative Zone," *Mideast Stars and Stripes*, October 10, 2009; Tom Coghlan, "Nato Commander Warns Of Conflict With Russia In Arctic Circle," *London Times*, October 3, 2009; Gerard O'Dwyer, "Danish Report: Conflicts Coming Over Arctic," *Defense News*, September 28, 2009: 18; Gerard O'Dwyer, "Russia Warns Denmark Over Arctic Arms Race," *Defense News*, August 3, 2009: 13; Gerrard Cowen, "Russia and NATO Look to Co-operation In Arctic," *Jane's Defence Weekly*, April 8, 2009: 12; David Scutro, "Thawing Arctic Seas May Bring Security Risk," *NavyTimes.com*, January 29, 2009.

[210] Department of Defense, *Quadrennial Defense Review Report*, February 2010, p. 86.

[211] Department of Defense, *Quadrennial Defense Review Report*, February 2010, p. 19.

[212] Department of Defense, *Quadrennial Defense Review Report*, February 2010, p. 57.

[213] Department of Defense, *Quadrennial Defense Review Report*, February 2010, p. 62.

[214] Jim Garamone, "Unified Command Plan Reflects Arctic's Importance," *American Forces Press Service*, April 7, 2011.

[215] The direction was contained in the House Armed Services Committee's report (H.Rept. 111-491 of May 21, 2010) on H.R. 5136, the FY2011 National Defense Authorization Act, which stated that the committee directs the Secretary of Defense to submit a report to the congressional defense committees by May 30, 2011, that includes the following:

(1) An assessment of the strategic national security objectives and restrictions in the Arctic Region;

(2) An assessment on mission capabilities required to support the strategic national security objectives and a timeline to obtain such capabilities;

(3) An assessment of an amended unified command plan that addresses opportunities of obtaining continuity of effort in the Arctic Ocean by a single combatant commander;

(4) An assessment of the basing infrastructure required to support Arctic strategic objectives, including the need for a deep-water port in the Arctic; and

(5) An assessment of the status of and need for icebreakers to determine whether icebreakers provide important or required mission capabilities to support Arctic strategic national security objectives, and an assessment of the minimum and optimal number of icebreakers that may be needed. (Page 337)

[216] Department of Defense, *Report to Congress on Arctic Operations and the Northwest Passage*, OUSD (Policy), May 2011. pp 2-4 (executive summary). See also Christopher J. Castelli, "Report: 'Significant' Gap Undermines DOD Tracking Of Ships In Arctic," *Inside the Pentagon*, June 16, 2011.

[217] The direction was contained on page 291 of in the House Armed Services Committee's report (H.Rept. 112-78 of May 17, 2011) on H.R. 1540, the FY2012 National Defense Authorization Act.

[218] Government Accountability Office, Arctic Capabilities[:]DOD Addressed Many Specified Reporting Elements in Its 2011 Arctic Report but Should Take Steps to Meet Near- and Long-term Needs, GAO-12-180, January 2012, Summary page and pp. 12-18.

[219] See Susan Gvozdas, "U.S. Coast Guard Preps For Open Arctic Waters," *Defense News*, September 7, 2009: 40; John C. Marcario, "Return to the Arctic," Seapower, August 2009: 26-27; John C. Marcario, "Coming Into Focus," Seapower, March 2009: 42, 44; John C. Marcario, "Arctic Presence," Seapower, August 2008: 32-33; Emelie Rutherford, "Coast Guard To Test How Ships and Aircraft Operate in Arctic," *Inside the Navy*, January 21, 2008; Philip Ewiing, "Allen: CG Expedition To Assess Arctic Waters," *NavyTimes.com*, January 17, 2009; Patricia Kime, "North Pole Flyover A First For Coast Guard," *NavyTimes.com*, October 25, 2007.

[220] Andrew Scutro, "Navy Preps For Uncharted Arctic Waters," *NavyTimes.com*, November 24, 2009.

[221] NSPD 66/HSPD 25, Section III B.

[222] Regarding infrastructure in the Arctic, an October 2009 press report stated:

A big challenge will be logistics support since the only [U.S.] supply base currently available is the town of Barrow, Alaska. [Rear Admiral Dave] Titley [the Oceanographer of the Navy] explained that the Arctic region is roughly the size of the United States. By way of illustration Titley said: "Think of trying to conduct operations through the entire U.S., and your one logistics base is the size of San Clemente Island [an island off the coast of Southern California]. That is roughly the size of Barrow."

(Bob Freeman, "Conference Addresses Navy's Role in a Changing Arctic," Navy New Service, October 4, 2009. See also Zachary M. Peterson, "As A Result of Melting Ice, Navy Study Need For Base in Far North," *Inside the Navy*, December 29, 2009; Patricia Kime, "CG Boss Calls For Forward Bases In The Arctic," *NavyTimes.com*, September 28, 2007.)

[223] For articles discussing potential changes to ship and aircraft designs, see Dan Taylor, "Roughead: Navy Will Have To Design Ships To Better Handle Arctic," *Inside the Navy*, November 26, 2009; Rebekah Gordon, "Coast Guard OPC [Offshore Patrol Cutter] Modifications For Varied Environments A Possibility," *Inside the Navy*, August 24, 2009; Andrew Scutro, "Arctic Forces Fleet to Look North," *NavyTimes.com*, February 23, 2009; Amy McCullough, "Stronger Hulls Could Help Fill Icebreaker Gap," *NavyTimes.com*, February 22, 2009; Rebekah Gordon, "Coast Guard Evaluating Effects of Increased Arctic Operations on Fleet," Inside the Navy, February 9, 2009.

[224] David Grove, "Arctic Melt: Reopening a Naval Frontier?" *U.S. Naval Institute Proceedings*, February 2009: 16-21.

[225] David W. Titley and Courtney C. St. John, "Arctic Security Considerations and the U.S. Navy's Roadmap for the Arctic," *Naval War College Review*, Spring 2010: 41.

[226] David W. Titley and Courtney C. St. John, "Arctic Security Considerations and the U.S. Navy's Roadmap for the Arctic," *Naval War College Review*, Spring 2010: 42.

[227] Lance M. Bacon, "Ice Breaker," *Armed Forces Journal*, March 2010: 16-19, 34-35.

[228] Source: Text of remarks of Chief of Naval Operations Admiral Gary Roughead at Active in the Arctic Seminar, June 16, 2011.

[229] Source: Text of remarks of Chief of Naval Operations Admiral Gary Roughead at the 4th Symposium on the Impacts of an Ice-Diminishing Arctic on Naval & Maritime Operations, June 20, 2011. See also Christopher J. Castelli, "Roughead: International Approach Needed To Fix Arctic-Awareness Gap," *Inside the Navy*, June 27, 2011.

[230] Nicole Klauss, "War Games Find U.S. Navy Ability Lacking In Acrtic," *Fairbanks Daily News-Miner*, April 28, 2012.

[231] Memorandum for Distribution dated November 10, 2009, from Admiral J. W. Greenert, Vice Chief of Naval Operations, on the subject of the Navy Arctic Roadmap. The document was posted on *InsideDefense.com* (subscription required).

[232] U.S. Navy, U.S. Navy Arctic Roadmap, Washington, 2009, 29 pp. (October 2009, sponsored by Task Force Climate Change [and] Oceanographer of the Navy.) The document was posted on *InsideDefense.com* (subscription required).

[233] For additional discussion of the roadmap, see David W. Titley and Courtney C. St. John, "Arctic Security Considerations and the U.S. Navy's Roadmap for the Arctic," *Naval War College Review*, Spring 2010: 35-48.

[234] Andrew Burt, "Navy Studies On Mission, Capability Needs For Arctic 'Nearing Completion'," *Inside the Navy*, January 21, 2011.

[235] Department of the Navy, *Arctic Environmental Assessment and Outlook Report*, August 2011, 25 pp. Accessed at http://greenfleet.dodlive.mil/files/2011/08/U.S.-Navy-Arctic-Environmental-Assessment.pdf.

[236] The POM is an internal DOD document that guides the development of DOD's annual budget submissions to Congress.

[237] Department of the Navy, *Arctic Environmental Assessment and Outlook Report*, August 2011, p. v.

[238] Nicole Klauss, "War Games Find U.S. Navy Ability Lacking In Acrtic," *Fairbanks Daily News-Miner*, April 28, 2012.

[239] Government Accountability Office, *Coast Guard[:]Efforts to Identify Arctic Requirements Are Ongoing, but More Communication about Agency Planning Efforts Would Be Beneficial*, GAO-10-870, September 2010, summary page.

[240] For additional discussion, see "Polar Icebreaking".

[241] *United States Coast Guard High Latitude Region Mission Analysis Capstone Summary*, July 2010, pp. 10-11, 13-15.

[242] Christopher J. Castelli, "DOD, Canada Aim TO Develop Arctic Policy Options By Early 2010," *Inside the Navy*, December 7, 2009.

[243] Christopher J. Castelli, "American, Canadian Defense Officials Tighten Ties On Arctic Issues," *Inside the Navy*, May 3, 2010.

[244] NORAD news release, "NORAD, USNORTHCOM and Canadian Joint Operations Commmand Commander Sign Cooperative U.S./Canada Documents," December 11, 2012, accessed December 21, 2012, at http://www.norad.mil/ News/2012/121112a.html.

[245] CRS communication with State Department official, October 8, 2010.

[246] On this question, a January 19, 2009, press article about the new directive stated:

> The new policy directive covers several key areas, including national security, energy exploration and the environment, but it does not specify whether any should take precedence over others.
>
> That led Jeremy Rabkin, a professor at George Mason University Law School, to comment: "It's really a list of all the things we're concerned about; that's not policy. I don't see anything here that helps you decide what gets priority."
>
> (Juliet Eilperin and Spencer S. Hsu, "White House Directive Guides Policy On Arctic," *Washington Post*, January 19, 2009: 2.)

[247] Title II of P.L. 98-373 is the National Critical Materials Act of 1984.

[248] Source for text: http://fas.org/irp/offdocs/nspd/nspd-66.htm. The text is also available online at http://www.nsf.gov/ od/opp/opp_advisory/briefings/may2009/nspd66_hspd25.pdf.

INDEX

A

abuse, 74, 76
access, viii, 5, 12, 15, 18, 21, 27, 28, 29, 45, 46, 48, 49, 55, 58, 61, 62, 72, 77, 83, 84, 89, 94, 105, 108, 123, 125, 130, 135, 136, 138, 156, 157, 158
accounting, 56
acidic, 13
acidity, 9
AD, 8
adaptability, 62
adaptation(s), 6, 21, 22, 30, 32, 45, 46, 57, 61, 62, 63, 64, 65, 72, 73, 74, 112, 145
adjustment, 56
advancement, 41, 42
adverse effects, 72, 160
adverse weather, 102
advocacy, 115, 129
aerosols, 4, 5, 35, 36, 37, 38, 39, 40, 42, 43, 47
Afghanistan, 128
age, 16, 68
agency collaboration, 1
Air Force, 127, 131, 140
air quality, 7, 31, 68, 69, 70, 71
air temperature, 4, 35, 97
airports, 29
Alaska Natives, 18, 68, 71, 73, 75, 76, 77, 113
Alaskan North Slope, 163
alcohol consumption, 76
alcoholism, 67, 68
algae, 59, 166
algorithm, 43
alternative energy, 46, 159
anthropology, 11
antibody, 73
appropriations, 160, 164
ARC, 93, 94
Arctic communities, 2, 6, 12, 29, 32, 33, 62, 63, 64, 65, 66, 72, 94, 158, 159
ARM, 39, 42, 48, 52, 81, 83, 85
armed conflict, 161
Asia, 92, 100, 101, 113, 137, 163, 169
assessment, 6, 10, 15, 22, 24, 37, 39, 56, 62, 66, 105, 116, 117, 130, 131, 141, 143, 144, 150, 164, 170, 171
assessment tools, 62
assets, 15, 35, 38, 77, 78, 82, 126, 127, 129, 140, 141, 142, 143, 145, 160
assimilation, 15, 45
atmosphere, 4, 8, 11, 13, 16, 18, 27, 31, 35, 36, 37, 38, 39, 41, 42, 43, 47, 53, 55, 56, 59, 60, 80, 81, 82, 97, 162
atmospheric stability, 4, 35, 47
authority(ies), 111, 121, 138, 140, 151, 154, 160, 164, 168
autonomy, 161
awareness, 79, 119, 128, 130, 131, 133, 134, 136, 140, 142, 144, 153, 154

B

bacterial infection, 68
balance of payments, 145
ballistic missiles, 125
barriers, 28, 114, 135
base, 20, 51, 61, 72, 93, 103, 127, 135, 171
BEA, 167
behavioral problems, 115
behavioral sciences, 75, 146, 147, 149
Beijing, 125
benchmarking, 52
benefits, 73, 75, 128
benign, 109
bilateral relationship, 128
biodiversity, 19

biological responses, 29
biological sciences, 149
biomass, 31
biosphere, 11
birds, 11, 109, 166
blood, 73
blueprint, vii, 2
boreal forest, 31
bottom-up, 61
breeding, 24
Broadband, 83, 84
Bureau of Land Management, 11
burn, 31, 102
businesses, vii, 2, 13

C

Cabinet, 103, 124
calibration, 45
campaigns, 5, 27, 36, 37, 38, 39, 41, 42, 43, 49, 52, 53, 55, 57, 58, 59
cancer, 7, 67, 68, 69, 70, 71
cancer screening, 68
carbon, 5, 16, 23, 29, 31, 36, 37, 38, 47, 53, 54, 59, 97, 124
carbon dioxide, 16, 23, 36, 47, 97
carbon monoxide, 36
Caribbean, 128
case studies, 62
cash, 62
casting, 13
CDC, 7, 10, 66, 67, 70, 71, 73, 74, 75, 76, 85
Census, 80
challenges, vii, viii, 2, 3, 8, 48, 59, 61, 62, 63, 64, 65, 70, 89, 94, 104, 106, 107, 108, 109, 110, 126, 129, 132, 135, 136, 138, 142, 144, 158, 160, 161
chemical, 4, 21, 37, 38, 54, 74, 80
chemicals, 71
Chicago, 169
child abuse, 67, 68
children, 69, 70, 76, 77
China, 19, 81, 88, 100, 123, 124, 125, 127, 169, 170
chronic diseases, 75
CIA, 25, 51
circulation, 12, 15, 17, 26, 27, 36, 46, 55, 57, 59
cirrhosis, 68
CIS, 169, 170
cities, 119
citizens, 45, 112, 143
civil service, 149
classes, 71
cleanup, 109, 110
climates, 145
clothing, 113, 139
clubbing, 112
CNN, 164
CO_2, 31
Coast Guard, viii, 5, 77, 79, 80, 81, 85, 87, 88, 90, 103, 104, 109, 115, 116, 117, 118, 119, 120, 121, 127, 128, 129, 130, 131, 132, 133, 134, 135, 137, 139, 141, 142, 143, 151, 163, 164, 165, 168, 170, 171, 172
coastal communities, 14, 64
Cold War, 125, 127, 135, 170
collaboration, 1, 2, 6, 7, 8, 15, 20, 28, 41, 44, 45, 51, 54, 59, 62, 63, 64, 65, 66, 67, 68, 70, 75, 83, 123, 131, 132, 133, 134, 144, 156
color, 21
combined effect, 164
commerce, 94, 113, 128, 136, 143, 154, 157
commercial, viii, 19, 22, 69, 70, 71, 77, 89, 90, 103, 111, 112, 114, 117, 118, 123, 124, 135, 139, 142, 143, 145, 157, 158, 160, 163, 164
commercial ties, 123
commodity, 101, 102, 107
communication, 1, 55, 65, 66, 75, 76, 82, 83, 102, 134, 136, 137, 161, 172
communication systems, 134
compatibility, 66
compensation, 147, 149, 152
competition, 90, 127, 129, 166
competitive advantage, 136
complement, 21, 35
complex interactions, 35, 54
complexity, 40
compliance, 70, 103
composition, 35, 37, 38, 39, 45, 159
compounds, 71
computer, 56, 61
computer systems, 56
computing, 58, 61, 136
conceptualization, 64
conference, 23, 115, 168, 169
conflict, 90, 126, 127, 129, 144
Congress, v, 12, 89, 90, 97, 103, 111, 114, 115, 116, 117, 128, 129, 142, 145, 148, 150, 151, 152, 161, 165, 168, 171, 172
Congressional Budget Act of 1974, 151
connectivity, 83, 84
consensus, vii, 3, 69, 70, 126, 141
consent, 96, 148, 156
conservation, 20, 24, 95, 111, 112, 156, 160
conserving, 10, 71
consolidation, 155

constituents, 5, 35, 36, 81
Constitution, 153
construction, 104, 107, 116, 118, 130, 135
consumers, 65
consumption, 71
containers, 102
contamination, 71, 109, 114
Continental, viii, 89, 96, 99, 106, 156, 161, 164
contingency, 70
convention, 95, 96, 138
cooling, 47, 97
cooperation, viii, 8, 9, 10, 12, 28, 44, 71, 81, 90, 95, 100, 119, 120, 121, 123, 124, 125, 126, 127, 136, 138, 144, 145, 146, 148, 150, 151, 152, 154, 157, 158, 159, 160, 161
cooperative agreements, 141, 158
coordination, vii, 1, 2, 6, 8, 18, 21, 28, 30, 32, 41, 42, 51, 53, 74, 77, 120, 121, 122, 128, 130, 134, 138, 144, 146, 148, 150, 154, 155, 156, 157, 158, 159, 160
correlation, 76
cost, 15, 32, 43, 53, 63, 68, 77, 102, 107, 114, 130, 143, 146
cost of living, 63
Court of Appeals, 107
crises, 119
critical infrastructure, 142, 144, 154
crude oil, 108, 110, 165
cryosphere, 8, 16, 29, 30, 51
cryptosporidium, 73
CSA, 49, 85
cultural heritage, 65, 66
culture, 73
customers, 102
cycles, 31, 43, 59
cyclones, 97
Cyprus, 103

D

data analysis, 26, 49
data collection, 15, 55, 59, 64
data communication, 143
data processing, 48, 84
data set, 4, 5, 35, 40, 48, 57, 58, 59, 60, 83
data transfer, 48
database, 33, 52, 62, 73
death rate, 6
decision makers, 133
deepwater drilling, 108
defence, 125, 144
deficiencies, 38
deformation, 11
degradation, 22, 31, 110
Delta, 16, 92
Denmark, viii, 8, 76, 88, 89, 91, 99, 100, 114, 120, 123, 125, 139, 154, 161, 163, 166, 170
Department of Agriculture, 6
Department of Commerce, 11, 85, 88, 111, 134, 149
Department of Defense, 3, 85, 118, 121, 128, 134, 139, 149, 170, 171
Department of Education, 85
Department of Energy, 3, 85, 149
Department of Health and Human Services, 7, 85, 149, 167
Department of Homeland Security, 85, 88, 117, 128, 129, 133, 134
Department of the Interior, 3, 8, 85, 149
Department of Transportation, 85, 149
deployments, 134
deposition, 36, 59
deposits, 36, 100, 104, 105, 106
depression, 67, 68
depth, 24, 37, 106, 128
detection, 32, 72, 73
deterrence, 134, 154
developed countries, 103
DHS, 5, 44, 49, 85, 131, 132, 133, 134
diabetes, 7, 114
diet, 73
digital technologies, 66
diplomacy, 127
diplomatic engagement, 129
direct measure, 37
disaster, 78, 103, 134, 138, 140, 158
disaster relief, 78, 158
discrimination, 39
diseases, 6, 7, 67, 69, 70, 72, 73, 76, 115
disorder, 67, 138
disposition, 117, 153, 164
distress, 101, 118, 122, 168
distribution, 18, 23, 24, 26, 31, 32, 37, 55, 59, 61, 78, 83, 114, 160
distribution function, 37
diversification, 61
diversity, 22, 38, 53
DOC, 11, 73, 85
DOI, 3, 4, 5, 6, 10, 11, 13, 14, 17, 24, 28, 30, 31, 32, 33, 44, 48, 49, 51, 52, 53, 55, 56, 57, 59, 60, 61, 66, 73, 79, 80, 83, 85
DOT, 85
draft, 82, 101, 103, 111, 130, 133
drainage, 25
drug abuse, 68, 76
Duma, 99

E

early warning, 72, 134, 136, 154
echinococcus, 73
ecological systems, 6
ecology, 18
economic activity, 108
economic change, 6, 65, 67, 113
economic consequences, 70, 159
economic cooperation, 126
economic development, vii, 1, 7, 10, 19, 154
economic growth, 114
economic indicator, 64
economics, 113
ecosystem, 4, 13, 14, 15, 17, 19, 20, 21, 22, 23, 24, 25, 28, 29, 30, 31, 33, 45, 47, 53, 57, 60, 61, 64, 72, 109, 110, 111, 124, 160, 166
education, 6, 65, 78, 83, 94, 132
educators, 49, 71, 94
egg, 112
embassy, 125
emergency, 78, 168
emergency response, 78
employment, 63, 103, 113
employment opportunities, 63
empowerment, 63, 100
endangered, viii, 90, 111, 112
endangered species, viii, 90, 111
endurance, 55, 143
end-users, 156
energy, vii, 2, 3, 5, 9, 12, 32, 33, 39, 42, 47, 49, 54, 71, 90, 94, 95, 106, 113, 124, 127, 134, 145, 146, 155, 156, 158, 159, 164, 172
energy efficiency, 71
energy security, 94, 156, 158
enforcement, 103, 160
engineering, 10, 73, 152
enlargement, 96
environment(s), vii, 1, 4, 7, 9, 10, 11, 12, 14, 21, 24, 26, 27, 29, 32, 42, 44, 45, 46, 70, 71, 72, 95, 98, 109, 110, 113, 123, 124, 126, 128, 130, 133, 134, 136, 138, 141, 142, 153, 157, 158, 159, 160, 166, 172
environmental awareness, 131
environmental change, vii, 2, 6, 8, 11, 17, 29, 30, 32, 46, 61, 62, 64, 94, 142
environmental conditions, 20, 27, 129
environmental factors, 68
environmental impact, 11, 107
environmental issues, 8, 10, 154
environmental protection, 95, 100, 101, 103, 123, 124, 131, 155, 156, 160
Environmental Protection Agency, 5, 86, 149, 160
environmental regulations, 107
environmental standards, 158
environmental stress, 32
EPA, 5, 6, 7, 44, 49, 61, 64, 66, 67, 73, 86
epidemiologic, 69, 70
epidemiologic studies, 70
epidemiology, 73
equipment, 17, 77, 79, 102, 104, 107, 109, 137, 139, 141, 142, 163
erosion, 30, 32, 48, 114
ethnic groups, 167
EU, 169
Eurasia, 16, 113
Europe, 42, 97, 100, 111, 128, 137, 169
European Parliament, 169
European Union, 99, 123, 135
evaporation, 110
evidence, 13, 16, 25, 27, 31, 35, 47, 104
evolution, 11, 34, 57, 59
execution, 43
executive branch, 90, 118
exercise, 45, 61, 98, 124, 126, 132, 140, 154, 156
expertise, 27, 48, 73, 147
exploitation, 155
exports, 124
exposure, 72, 110, 114
extinction, 112, 166
extraction, 69, 97, 101, 108, 114, 137
extreme cold, 145
extreme weather events, 76

F

Fairbanks, 50, 77, 165, 168, 171, 172
families, 65, 69, 113, 166
fast processes, 57
fauna, 72
federal agency, 93, 108
federal assistance, 114
federal government, 117
Federal Government, 146, 148, 150
Federal Register, 147, 166
fiber, 61
financial, 149
Finland, 8, 67, 88, 91, 92, 113, 120, 154, 166, 169
fires, 4, 29, 31, 32, 36, 37, 71, 97
fish, viii, 11, 79, 90, 111, 113, 145
Fish and Wildlife Service, 86, 112, 166
fisheries, viii, 90, 111, 141, 160, 161

fishing, 11, 69, 70, 71, 97, 100, 104, 111, 112, 113, 114, 137, 138, 160
flexibility, 102
flight(s), 118, 135, 136
flooding, 32, 114
floods, 72
flora, 72
flotation, 71
fluctuations, 65
focus groups, 71
food, 65, 71, 72, 73, 97, 113, 114
food safety, 114
food web, 71
force, 33, 95, 96, 106, 117, 139, 141, 143, 156, 161
forecasting, 4, 13, 14, 15, 17, 29, 49, 51, 128, 129
foreign investment, 114
foreign policy, 125, 150, 163
forest fire, 36
formation, 5, 11, 26, 37, 38, 40, 41, 59
France, 163, 169, 170
freedom, 127, 134, 135, 136, 138, 154
freezing, 55, 137, 141
freshwater, 26, 46, 49, 114
funding, 3, 22, 23, 41, 59, 64, 74, 75, 115, 116, 118, 134, 141, 142, 151
funds, 59, 74, 83, 94, 139, 149, 151

G

GAO, 32, 86, 110, 114, 130, 141, 165, 167, 171, 172
gastrointestinal tract, 76
GDP, 125
General Services Administration, 149
Geographic Information System, 30, 86
geography, 126
geology, 11, 78
geometry, 26, 27, 56
Germany, 78
GHG, 97
giardia, 73
GIS, 30, 86
glacial-process studies, 4
global climate change, viii, 89, 139
global mobility, 154
global warming, 47, 97, 99, 139, 162
globalization, 113
governance, 95, 110, 123, 155, 161
governments, 45, 46, 49, 77, 90, 99, 113, 115, 128, 155, 156, 160
GPS, 82, 86
grants, 94, 150
gravity, 56
greenhouse, 8, 16, 17, 36, 47, 97
greenhouse gas(es), 8, 16, 17, 36, 97
grounding, 56
growth, 15, 22, 116
guidance, 84, 129, 140, 150
guidelines, viii, 69, 89, 101, 104, 121, 150, 168
Gulf Coast, 108
Gulf of Mexico, 108, 119, 164, 165

H

habitat(s), 25, 29, 30, 112, 146
harvesting, 61, 63, 65, 66, 114
hazards, 67, 70, 72, 111, 136, 144, 158
health, vii, viii, 2, 3, 6, 7, 9, 45, 46, 62, 63, 67, 68, 69, 70, 71, 72, 73, 74, 75, 76, 90, 113, 114, 146, 147, 155, 167
health effects, 72
health problems, 68, 74, 114
health risks, 114
heart disease, 67
height, vii, 2, 21, 39
helicobacter pylori, 76
Helicobacter pylori, 67, 68, 69, 70
hemisphere, 92
hepatitis, 67, 68, 69, 70
HHS, 7
history, 11, 13, 23, 108, 110, 136
homeland security, 7, 93, 135, 136, 153, 154
homes, 68, 70, 71
host, 69, 70, 81, 106, 166
hostile acts, 154
hot spots, 19
hotel, 119
House, 12, 119, 170, 171
housing, 68
hub, 28, 42
human, vii, viii, 1, 2, 6, 7, 11, 12, 18, 19, 20, 22, 24, 30, 32, 36, 46, 48, 62, 67, 72, 76, 89, 90, 94, 97, 108, 109, 110, 112, 119, 120, 122, 123, 128, 129, 130, 141, 142, 145, 153, 155, 158, 159, 160, 168
human actions, 112
human activity, 11, 18, 32, 36, 108, 110, 119, 120, 123, 129, 130, 141, 142, 153, 155, 158, 159, 160
Human Development Report, 61, 161, 166
human dimensions, 2, 11
human health, 7, 12, 36, 46, 67, 72, 73, 76, 145, 160, 168
humanitarian aid, 158
hunting, 11, 113, 114

hydrocarbons, 110
hydrosphere, 11

I

ice pack, 99
Iceland, 8, 28, 51, 67, 88, 91, 97, 111, 120, 125, 154, 161, 166, 169
ICS, 67, 69, 86
ideal, 60
identification, 21, 33, 37, 56, 144
illumination, 39
image(s), 102, 164
imagery, 49
immune function, 72
impact assessment, 57, 64, 159
Impact Assessment, 25, 33, 51, 76, 84, 166
imports, 125
improvements, 14, 22, 53, 54, 61, 148
incidence, 68, 69
independence, 161
India, 100, 123, 169
Indians, 77, 167
indigenous knowledge, 115
indigenous peoples, viii, 6, 67, 69, 70, 90, 112, 113, 114, 115, 167
indirect effect, 36, 39
individuals, 3, 45, 46, 147, 166
industry(ies), 14, 15, 17, 21, 46, 49, 69, 70, 94, 103, 107, 147
infant mortality, 6, 67
infants, 68, 72
infection, 68, 72, 73
infectious agents, 74
information exchange, 28, 75, 76, 100
information sharing, 27, 28, 127
information technology, 74
infrastructure, 8, 14, 17, 25, 29, 32, 61, 62, 63, 77, 78, 82, 102, 103, 104, 105, 107, 110, 114, 119, 129, 130, 131, 132, 134, 135, 136, 142, 143, 157, 159, 165, 170, 171
injury(ies), 6, 7, 67, 68, 72, 147, 153
insecurity, 114
institutions, 8, 10, 11, 65, 74, 75, 82, 148, 149, 154, 157
integration, 18, 22, 23, 48, 49, 52, 57, 58, 60, 61, 84
integrity, 32, 125, 143
intelligence, 129, 144
Interagency Arctic Research Policy Committee (IARPC), v, vii, 1, 2, 3, 7, 93, 157
interagency coordination, vii, 3, 43, 53, 150, 151
interface, 26, 27, 46, 56
intermediaries, 41
international law, 129, 153, 156, 164
International Maritime Organization (IMO), 103, 104, 110, 111, 155, 157
international standards, 158
international trade, 103
interoperability, 48, 139
intervention, 69, 70
intervention strategies, 69, 70
investment(s), 54, 93, 94, 102, 107, 125, 129, 130, 131, 132, 133, 134, 141, 143, 144, 157
Iraq, 165
Iraq War, 165
iron, 104
islands, 12, 138
isolation, 145
issues, viii, 8, 20, 35, 61, 66, 74, 90, 95, 100, 106, 111, 117, 121, 123, 124, 126, 127, 128, 133, 139, 142, 155, 156, 158, 159, 165

J

Japan, 15, 19, 41, 81, 88, 100
jurisdiction, 95, 96, 120, 142, 154, 159, 160

K

Kola Peninsula, 166
Korea, 19, 81, 88, 163, 170

L

laboratory studies, 40
lakes, 26, 71
landings, 79
landscape, 31
language skills, 65
languages, 65, 66, 113, 166
larvae, 109
law enforcement, 115, 154
Law of the Sea Convention, 96, 156
laws, 96, 107, 142, 149, 153, 168
laws and regulations, 107, 168
lead, viii, 4, 14, 15, 17, 22, 25, 29, 32, 38, 47, 53, 67, 71, 72, 88, 89, 93, 94, 96, 112, 114, 115, 122, 130, 131, 146, 149, 164
leadership, 10, 93, 94, 129, 157
legislation, 111
Liberia, 103
lichen, 31
life cycle, 112
life expectancy, 67, 113

lifetime, 37
light, 37, 38, 79, 118, 138
light scattering, 37
liquids, 105
litigation, 107
liver, 68
local community, 5, 46, 49
local government, 65, 114, 146, 148, 150
logistics, 50, 94, 136, 142, 158, 171
longevity, 5
lying, 38

M

machinery, 102
magnitude, 26, 35, 47, 143
majority, 67, 96, 147
mammal(s), 11, 18, 20, 24, 79, 109, 112, 113, 114, 146, 166
man, 71, 125
management, viii, 11, 14, 17, 20, 25, 30, 31, 32, 61, 63, 64, 66, 69, 84, 90, 111, 112, 145, 155, 156, 157, 158, 159, 160
manpower, 139
mantle, 56
manufacturing, 100
mapping, 30, 37, 78, 104, 132, 134
Marine Corps, 141
marine environment, 5, 16, 21, 47, 72, 112, 157, 158
marine fish, 114
Marine Mammal Protection Act, 112
marine species, 156
maritime security, 134, 136, 154
Marshall Islands, 103
mass, vii, 2, 3, 6, 9, 15, 18, 22, 25, 34, 35, 38, 45, 46, 49, 55, 56, 59, 78, 82, 145
mass loss, 6, 15, 25, 34, 35, 45, 46, 56
materials, 65, 146, 149
matter, 120
measurement(s), 15, 19, 21, 25, 26, 27, 37, 38, 39, 40, 41, 42, 43, 46, 77, 79, 80
media, 65, 66, 78, 164
medical, 73
melt, viii, 15, 26, 27, 38, 59, 89, 97, 99, 104, 124
melting, 15, 18, 25, 27, 45, 55, 57, 71, 94, 97, 100, 101, 102, 161
melts, 105, 107
membership, 147
meningitis, 69
mental health, 7, 74
mental illness, 74, 114
mercury, 36, 72, 114, 160
messages, 126
Mexico, 128
micronutrients, 72
migration, 29, 65, 166
militarization, 126
military, viii, 12, 90, 100, 104, 118, 123, 124, 125, 126, 127, 128, 135, 136, 137, 139, 144, 154
military exercises, 144
mineral resources, 126
Minerals Management Service (MMS), 165
minorities, 113, 167
mission(s), 11, 15, 16, 41, 43, 44, 49, 77, 78, 79, 81, 115, 116, 117, 119, 121, 127, 131, 133, 134, 135, 136, 138, 140, 142, 143, 144, 164, 170, 171
models, vii, 2, 3, 4, 5, 6, 9, 13, 14, 15, 16, 17, 18, 20, 21, 22, 23, 26, 27, 32, 35, 36, 37, 38, 39, 40, 41, 42, 43, 45, 46, 47, 49, 51, 52, 53, 54, 55, 56, 57, 58, 59, 60, 61, 64, 65, 70, 97, 136, 162
Moderate Resolution Imaging Spectroradiometer (MODIS), 36, 86
modernization, 62, 67, 117
modifications, 136
MODIS, 37, 39, 86
moisture, 16, 40, 78
Montreal Protocol, 155
Moon, 25, 34
moratorium, 112, 164
morbidity, 7, 69, 70, 72
morphology, 59
mortality, 7, 67, 68, 69, 70, 72
mortality rate, 67, 68
Moscow, 100, 120, 122, 123, 126, 128
Multilateral, 123
mutation, 166
mutilation, 112

N

NAS, 50
nation states, 129
National Academy of Sciences, 45, 162, 165
National Aeronautics and Space Administration, 3, 86, 149
national borders, vii, 2, 124
National Defense Authorization Act, 133, 170, 171
National Institute for Occupational Safety and Health, 67, 87, 88
National Institutes of Health, 7, 87

national interests, 96, 127, 129, 130, 136, 138, 142, 156
National Park Service, 66, 87, 88
national policy, vii, 2, 3, 10, 93, 117, 142, 146, 150
National Research Council, 50, 108, 138, 165
national security, vii, viii, 1, 7, 10, 14, 25, 77, 89, 95, 97, 103, 106, 129, 130, 132, 134, 135, 136, 153, 154, 155, 170, 171, 172
national strategy, 134
National Strategy, 160
nationality, 122, 168
Native communities, 46, 62
native population, 75
NATO, 123, 126, 128, 169, 170
natural gas, 22, 23, 104, 105, 156
natural resource management, 7, 10, 154
natural resources, vii, 3, 25, 95, 101, 108, 123, 124, 146, 149, 155, 156, 158, 159, 160
natural science(s), 94
natural selection, 166
navigation system, 143
negative outcomes, 6
neglect, 56, 146
negotiating, 99
negotiation, 123
New Zealand, 163
NOAA, 3, 5, 6, 10, 11, 13, 14, 15, 17, 18, 19, 20, 21, 23, 24, 27, 30, 34, 35, 37, 38, 39, 40, 41, 42, 43, 44, 45, 48, 49, 51, 52, 53, 54, 55, 56, 57, 58, 59, 60, 61, 65, 73, 77, 78, 79, 80, 81, 83, 87, 88, 103, 111, 162, 163, 164
nodes, 82
North America, 12, 69, 92, 100, 101, 113, 118, 128, 144, 167
North American Free Trade Agreement, 69
Norway, viii, 8, 44, 67, 88, 89, 91, 98, 100, 113, 120, 122, 123, 124, 125, 126, 127, 154, 161, 162, 163, 166, 168, 169, 170
NPS, 87
NRC, 108, 165
nucleation, 38, 60
nutrients, 18, 22, 59
nutrition, 76, 113

O

Obama, 95, 128, 129, 145
Obama Administration, 95, 145
obesity, 7, 114
obstacles, 110, 114
oceans, 8, 9, 45, 54, 95, 114, 155
OCS, 105, 107, 164, 165
Office of Management and Budget, 151
officials, 125, 127, 128, 131, 132, 139, 141, 144
oil, viii, 11, 17, 21, 22, 23, 69, 70, 72, 89, 90, 96, 97, 98, 100, 103, 104, 105, 106, 107, 108, 109, 110, 112, 118, 119, 120, 124, 127, 135, 143, 145, 156, 158, 159, 163, 164, 165
oil spill, viii, 17, 23, 70, 72, 89, 108, 109, 110, 120, 124, 127, 143, 165
Omnibus Appropriations Act,, 164
operating costs, 102
operations, 12, 14, 17, 45, 49, 71, 79, 80, 88, 90, 103, 105, 108, 110, 116, 118, 120, 121, 122, 126, 127, 128, 129, 131, 132, 133, 134, 135, 136, 137, 138, 139, 140, 142, 143, 144, 151, 154, 168, 170, 171
opportunities, vii, 2, 12, 23, 52, 54, 64, 66, 83, 94, 97, 106, 110, 114, 117, 124, 128, 129, 130, 131, 132, 133, 138, 144, 146, 157, 159, 170
optical properties, 37
organ, 21
organize, 13, 146
Outer Continental Shelf (OCS), 8, 11, 105, 164, 165
Outer Continental Shelf Lands Act, 164
outreach, 6, 66, 83
overlap, 132, 133, 159
oversight, 145
ownership, 83, 103, 114, 116
oxidation, 36
ozone, 36, 37, 47

P

Pacific, 12, 19, 20, 21, 23, 47, 67, 87, 88, 97, 111, 112, 129, 135, 139, 164, 169
Panama, 103, 163
parallel, 23, 56, 96
Parliament, 163
participants, 20, 44, 75, 115, 123, 125, 169
particle nucleation, 40
pathogenesis, 69, 70
pathways, 62
PCBs, 87
peace, 161
peacekeeping, 161
peer review, 75
Pentagon, 129, 171
peptic ulcer, 68, 69, 70
peptic ulcer disease, 68, 69, 70
permafrost, viii, 4, 5, 8, 9, 16, 22, 29, 30, 31, 32, 36, 45, 47, 48, 49, 53, 54, 57, 59, 71, 89, 92, 97, 104, 107, 109, 114
permission, 75, 121

permit, 99, 124, 128
Persistent Organic Pollutants, 87
petroleum, 100, 105, 107, 125
Petroleum, 165
physical environment, 9, 20, 136
physical properties, 5, 38
physicians, 71
physics, 17, 56
pipeline, 53, 163
plankton, 11, 18, 20
plants, 60, 112, 166
platform, 32, 67, 151
pleasure, 104
pneumonia, 69
polar, viii, 4, 25, 35, 43, 47, 78, 80, 90, 93, 94, 97, 102, 104, 111, 112, 115, 116, 117, 118, 122, 125, 126, 138, 142, 143, 145, 166, 168
policy, viii, 6, 8, 10, 14, 17, 31, 32, 61, 64, 65, 69, 70, 84, 89, 93, 94, 95, 97, 109, 111, 115, 118, 123, 136, 141, 145, 146, 148, 149, 150, 151, 153, 154, 155, 156, 157, 158, 159, 160, 163, 172
policy issues, 115
policy makers, 6
policymakers, vii, 2, 94
political leaders, viii, 90
pollutants, 6, 67, 72, 160
pollution, viii, 36, 39, 89, 97, 103, 104, 108, 109, 114, 145, 158
polychlorinated biphenyls (PCBs), 71
ponds, 59
pools, 47
POPs, 72, 87
population, 33, 65, 67, 68, 72, 75, 94, 109, 112, 114, 126, 137, 138, 166
population size, 33
portfolio, 94
positive feedback, 47
power generation, 164
precedent, 96
precipitation, 12, 32, 38, 40, 71
prediction models, 5, 40, 41, 42
prejudice, 96, 120
preparation, 84, 144, 150
preparedness, 139
preservation, 6, 65, 66
President, 2, 10, 93, 100, 125, 126, 128, 146, 147, 148, 149, 150, 151, 152, 164, 165, 169
President Obama, 10, 125
presidential directive, viii, 89, 95, 106, 153
Presidential Directives, 164
prevention, 67, 73, 76, 103, 104, 158
price stability, 107
Prince William Sound, 165
principles, 26, 138, 158, 160
private sector, 49, 108, 150
procurement, 149, 151
professionals, 72, 109
project, 15, 20, 21, 22, 32, 34, 51, 59, 68, 69, 70, 72, 97, 116, 154, 163
protection, 20, 115, 135, 142, 144, 158, 160
public health, 67, 73, 114
public interest, 150

Q

quality control, 58
quantification, 4, 60

R

race, 98, 123, 126, 138
radar, 16, 39, 49
radiation, 5, 16, 36, 38, 39, 41, 79
Radiation, 39, 47, 48, 81, 85
rate of change, 29, 112
reactions, 113
reality, 102
recognition, 23, 84, 113, 156
recommendations, 29, 33, 44, 61, 64, 68, 73, 96, 104, 111, 115, 129, 131, 133, 140, 148, 150, 151, 153, 156
reconstruction, 116
recovery, 84, 109, 110, 143, 146
recycling, 116
redistribution, 114
reflectivity, 4
regional policy, 129
registry(ies), 68, 69, 70, 71, 103
regrowth, 15
regulations, 101, 103, 104
remote sensing, 15, 17, 25, 27, 35, 40
renewable energy, 105
repair, 115
requirements, 15, 23, 27, 84, 101, 104, 110, 112, 116, 117, 131, 133, 134, 138, 140, 141, 142, 149, 151, 158, 160
RES, 168
research institutions, 147, 157
researchers, 8, 33, 48, 59, 66, 75, 76, 81, 83, 94
reserves, 98, 118
resilience, 6, 20, 24, 61, 64, 66
resistance, 110
resolution, 6, 18, 36, 38, 39, 53, 54, 55, 57, 58, 61, 136, 155
resource allocation, 67

resource management, 20, 156
respiratory syncytial virus, 67, 68, 76
response, 4, 16, 17, 20, 23, 25, 29, 31, 32, 44, 56, 69, 70, 73, 103, 107, 109, 110, 111, 112, 114, 115, 119, 122, 126, 127, 138, 140, 143, 144, 157, 158, 165, 166
restrictions, 101, 105, 170
retirement, 166
rights, 46, 96, 100, 113, 114, 120, 154, 155, 156, 164
risk(s), viii, 12, 20, 23, 69, 70, 71, 72, 73, 76, 89, 97, 102, 105, 108, 109, 114, 118, 126, 130, 131, 132, 133, 139, 143, 157, 158, 160
risk assessment, 23, 143
risk factors, 69, 70
root, 109
root system, 109
routes, viii, 89, 100, 101, 102, 122, 124, 163
rules, 123
Russia, viii, 15, 16, 19, 35, 42, 80, 81, 88, 89, 91, 97, 98, 99, 100, 104, 106, 113, 115, 123, 124, 125, 126, 127, 135, 139, 156, 162, 163, 166, 167, 169, 170

S

safety, 67, 69, 70, 71, 77, 94, 97, 101, 103, 104, 113, 120, 129, 136, 140, 142, 143, 144, 158, 163
salinity, 21, 46
salmon, 72, 73
sample design, 75
savings, 102
scaling, 51
Scandinavia, 124
school, 65
science, vii, 2, 3, 6, 8, 11, 12, 18, 20, 23, 25, 31, 33, 35, 51, 58, 59, 60, 61, 64, 79, 81, 94, 137, 141, 144, 157, 164
scientific knowledge, 62, 142
scientific observation, 94
scientific understanding, 5, 157
scope, vii, 1, 3, 61, 126, 155, 168
sea level, 6, 9, 12, 25, 26, 30, 33, 34, 35, 55, 56, 94, 114, 138
seafood, 145
sea-level, 8, 25, 33, 34, 56
sea-level rise, 8, 34
seasonality, 46
Secretary of Commerce, 112
Secretary of Defense, 133, 170
security, viii, 73, 90, 103, 123, 124, 125, 126, 127, 128, 129, 134, 136, 138, 140, 142, 144, 146, 158
sediments, 59, 106
seed, 112
seismic data, 164
self-sufficiency, 63
Senate, 95, 99, 117, 119, 133, 155, 156
sensing, 4, 19, 35, 37, 39, 45
sensitivity, 47
sensors, 18, 21, 38, 39, 79, 82, 136
services, 4, 7, 14, 17, 24, 25, 29, 33, 53, 68, 69, 70, 71, 83, 103, 148, 149, 152
settlements, 100
sexually transmitted diseases, 67
shoreline, 114
shortage, 63, 119
showing, 93, 161
Siberia, 16, 18, 166
signals, 138, 164
signal-to-noise ratio, 27
signs, 38
simulation(s), 26, 37, 38, 39, 51, 53, 56, 57, 58, 60, 61, 83, 162
Singapore, 100
skin, 68, 76
Smithsonian Institution (SI), 4
smoke plume, 36
social development, 94, 100
social indicator, 33
social stress, 72
society, 30
software, 78
solidarity, 115
solution, 60, 116
solution space, 60
South Korea, 100
Southeast Asia, 100
sovereignty, 90, 96, 100, 106, 115, 120, 125, 126, 127, 135, 154, 163
Soviet Union, 100, 125, 127, 128, 145, 146
Spain, 169
specialists, 152
species, 14, 17, 18, 20, 22, 29, 30, 65, 71, 72, 73, 90, 108, 109, 112, 160, 166
speech, 170
spending, 125, 151
Spring, 120, 171, 172
stability, 4, 29, 35, 47, 54, 140
stakeholders, 1, 14, 44, 49, 57, 62, 63, 69, 109, 130, 131, 140
state control, 103
state of emergency, 168

statistics, 118
sterile, 166
storage, 23, 32, 105
storms, 20, 72, 102
stress, 26, 27, 56, 115, 128
stressors, 6, 61, 63, 64, 159
structure, 21, 39, 40, 41, 42, 47, 55, 113, 128, 141, 155
submarines, 77, 127, 130
subsistence, 14, 17, 18, 19, 20, 22, 23, 29, 30, 31, 32, 46, 65, 71, 72, 73, 74, 113, 114, 147, 152, 155, 159
substance abuse, 7, 67, 74
succession, 31, 32
suicide, 6, 7, 67, 68
surface properties, 38
surveillance, 7, 67, 69, 73, 76, 99, 129
survival, 18, 65, 67, 71, 97
survivors, 120
susceptibility, 72
sustainability, 62
sustainable development, 67, 68, 76, 87, 88, 124, 155
sustained development, 107
Sweden, 8, 67, 88, 91, 92, 113, 120, 124, 125, 154, 166, 169
symptoms, 69, 70
synthesis, 17, 22, 37, 40, 44, 57, 60, 88

T

target, vii, 2, 15
Task Force, 45, 62, 76, 120, 122, 137, 138, 139, 142, 172
teams, 49
technical support, 58
techniques, 27, 55, 57, 58, 60, 109
technological advances, 71
technological progress, 71
technology(ies), 8, 18, 19, 21, 22, 23, 27, 41, 66, 71, 73, 82, 105, 107, 128, 131, 143, 144, 146, 164
temperature, 15, 21, 32, 46, 47, 71, 92, 97, 109, 112, 137, 160, 162
tension, 126
terrestrial ecosystems, 14, 25, 27
territorial, viii, 89, 91, 96, 99, 106, 115, 125, 154, 161
territory, viii, 10, 21, 89, 91, 96, 99, 100, 107, 114, 121, 123, 125, 129, 142, 151, 161, 168
terrorism, 126, 154
terrorist attack, 154
testing, 17, 45, 56, 79, 142, 145, 164
thinning, 25, 47, 53
threat assessment, 144
threats, 24, 65, 66, 71, 72, 73, 107, 114, 126, 128, 135, 144, 161
time series, 72
tissue, 72, 73
Title I, 7, 91, 93, 145, 161, 172
Title II, 161, 172
top-down, 61
tourism, viii, 12, 89, 97, 108, 114, 143, 155
toxic metals, 72
toxicity, 109
trade, 100, 101, 112
trading bloc, 100
training, 72, 73, 104, 132, 134, 135, 136, 141, 144
trajectory, 94
transmission, 23, 78, 82
transport, 26, 35, 36, 37, 38, 39, 40, 41, 46, 59, 72, 97, 101, 104, 105, 107, 135
transportation, vii, 1, 14, 62, 78, 95, 103, 106, 107, 110, 123, 125, 136, 137, 146, 157, 158, 159
transportation infrastructure, 110
transshipment, 102
trauma, 7, 68, 72
treaties, 103, 153, 155
treatment, 67, 69, 70
triggers, 68, 69, 98
tuberculosis, 67, 69, 70
tularemia, 73
tundra, 4, 31, 34, 36, 45, 107, 114

U

U.S. Army Corps of Engineers, 103, 114, 167
U.S. coastal states, 104
U.S. Geological Survey, 8, 105, 138, 164
U.S. policy, viii, 89, 95, 135, 145, 153
underwater vehicles, 82
uniform, 68, 83, 162
unintentional injuries, 68
unique features, 4, 35
United Kingdom (UK), 33, 34, 43, 50, 76, 81, 169
United Nations, 95, 103, 115, 128, 155, 160, 168
universities, 149
unresolved Arctic territorial disputes, viii, 89, 99
updating, 155, 158
USA, 34, 43, 50
USDA, 6
USGS, 27, 28, 30, 34, 48, 64, 78, 83, 87, 105, 106, 164

UV, 82

V

validation, 6, 15, 38, 40, 41, 43, 46, 56, 60, 61
variables, 15, 16, 18, 21, 27, 45, 46, 78, 142
vector, 74
vegetation, 4, 30, 31, 32, 51, 53, 57, 60, 78, 97, 114
vehicles, 39, 42, 80, 136
ventilation, 137
venue, 107, 138, 155
vessels, 18, 21, 77, 80, 81, 100, 101, 102, 103, 104, 110, 116, 117, 118, 124, 125, 129, 130, 136, 142, 146, 154
Vice President, 163
virology, 69, 70
virus infection, 68
viscosity, 110
vision, 10, 44
visualization, 83
voters, 161
vulnerability, 62, 63, 64, 154, 159

W

Warsaw Pact, 125
Washington, 12, 13, 50, 52, 76, 116, 120, 122, 125, 161, 165, 167, 169, 172
waste, 71
water, 8, 16, 18, 22, 26, 27, 30, 39, 48, 59, 68, 69, 71, 73, 76, 78, 82, 101, 106, 110, 113, 115, 132, 135, 137, 138, 160, 170
water purification, 69
water resources, 48
water supplies, 71
waterways, 125, 158
wealth, 125, 126
weather patterns, 45, 94, 145
web, 28, 73, 111, 170
welfare, 45
well-being, vii, 1, 46, 61, 62, 65, 67, 124
whales, 18, 112
White House, 172
White Paper, 134, 165
wilderness, 96
wildfire, 31, 34
wildland, 4, 29, 31, 71
wildlife, 4, 7, 31, 48, 65, 72, 73, 74, 107, 109, 113, 115, 146, 155
wildlife conservation, 155
windows, 102
workers, 70, 72
working groups, 44, 124
worldwide, 42, 99, 115, 155

Z

zooplankton, 18